B. Fivaz T. Petney I. Horak

Tick Vector Biology

Medical and Veterinary Aspects

With 17 Figures and 9 Tables

Springer-Verlag
Berlin Heidelberg New York
London Paris Tokyo
Hong Kong Barcelona
Budapest

Prof. BRUCE FIVAZ
Tick Research Unit, Rhodes University
Grahamstown 6140, South Africa

Dr. TREVOR PETNEY
University of Heidelberg, Dept. of Parasitology
Im Neuenheimer Feld 324
D-6900 Heidelberg

Prof. IVAN HORAK
Veterinary Entomology, Faculty of Veterinary Science
University of Pretoria
Pretoria, South Africa

The cover illustration depicts 3 male *Amblyomma variegatum*, one of the major vectors of the disease heartwater. This illustration follows an original by Andre Olwage in an article by J. B. Walker which appeared in the *Onderstepoort Journal of Veterinary Research.*

Library of Congress Cataloging-in-Publication Data.
Tick vector biology: medical and veterinary aspects / [edited by] B. Fivaz, T. Petney, I. Horak.
p. cm. Includes bibliographical references and index.
ISBN-13: 978-3-642-76645-9 e-ISBN-13: 978-3-642-76643-5
DOI: 10.1007/ 978-3-642-76643-5
1. Ticks as carriers of disease. 2. Tick-borne diseases in animals. 3. Tick-borne diseases. 4. Ticks. 5. Ticks – Control. 6. Ticks as carriers of diseases – Africa, Southern. 7. Tick-borne diseases in animals – Africa, Southern. 8. Tick-borne diseases – Africa, Southern. 9. Ticks – Africa, Southern. 10. Ticks – Control – Africa, Southern. I. Fivaz, B. (Bruce) II. Petney, T. (Trevor) III. Horak, I.
SF810.T5T53 1992 636.089'433–dc20 92-7684 CIP

Softcover reprint of the hardcover 1st edition 1992

Production Editor: Herta Böning, Heidelberg
Typesetting: Macmillan Ltd., Bangalore, India
31/3145-5 4 3 2 1 0 – Printed on acid-free paper

The Tick Research Unit, Rhodes University

The association between tick infestation and stock losses was recognized in the eastern Cape Province of South Africa in the latter half of the 19th Century. This resulted in a Cattle Disease Commission being appointed at Grahamstown in 1877.

The problem became so intense that tick control by means of dips and washes was initiated and sodium arsenite applied in a plunge dip became the control measure of choice.

Tick resistance to arsenic was noted towards the end of the 1930s and from then on also to the newer acaricides. This growing problem prompted Rhodes University and the Mohair Grower's Association of South Africa to arrange a conference at the University to discuss the future direction of tick research. The proceedings of this conference which was held in 1969, were published under the title *The Biology and Control of Ticks in Southern Africa.*

This conference prompted the establishment of the Tick Research Unit in the Department of Zoology at Rhodes University in 1971 with G. B. Whitehead as its first director. Upon his retirement I. G. Horak became director and he was followed by the present director B. H. Fivaz.

Since its establishment a number of prominent researchers have worked in the Tick Research Unit. These include J. G. H. Londt, R. A. I. Norval, D. R. Arthur, Y. Rechav and T. N. Petney. More than 100 scientific papers have been produced by the Unit in the 20 years of its existence.

On its 10th anniversary in 1981 the Unit arranged an international conference and the proceedings of this conference were published under the title of *Tick Biology and Control.*

To celebrate the 20th anniversary of the Tick Research Unit it was decided to publish a book on some of the latest topics in tick and tick-borne disease research. This publication is the outcome of that decision.

Grahamstown 1991

B. H. FIVAZ
T. N. PETNEY
I. G. HORAK

Contents

Systematics of the Ixodida (Argasidae, Ixodidae, Nuttalliellidae): An Overview and Some Problems

J.E. KEIRANS[1]

Summary

The world's tick fauna is composed of some 850 species in 3 families and 19 genera. All known tick species are obligate hematophagous ectoparasites of either warm- or cold-blooded animals during one or more stages of their life cycle. Interest in ticks has increased dramatically in recent years primarily due to the relationship between ticks of the *Ixodes ricinus* group and their ability to transmit the causative agent of Lyme disease to humans.

In this chapter, the world's tick fauna is listed; genera and occasionally subgenera considered; taxonomic problem areas discussed; and pertinent literature reviewed and cited.

1. Introduction

Ehrlich (1961) suggested that if Linnaeus were to materialize in the laboratory of a modern museum taxonomist, he would probably feel right at home. As one of the toilers in the taxonomic vineyard, I find that 30 years later this statement remains essentially true. Others will undoubtedly interpret this view as heretical and insist that recent developments in computerized expert systems, compact discs, phylogenetic systematics (i.e. cladistics), biochemical systematics including electrophoresis of soluble proteins, and molecular techniques such as species-specific DNA sequencing have revolutionized the classification of organisms. To a certain extent this is true, but even today, if Linnaeus were to enter most taxonomic laboratories in this or any other country, he would find taxonomists sorting, arranging, naming, and deciding where, within the scheme of classification, their organisms belong. And most taxonomists would not be using "cutting edge" techniques to create their classifications.

When dealing with species of insects or acarines of potential medical importance, the taxonomist is almost always called upon to identify the vector and then to tell whether or not the species in question is capable of transmitting a disease organism. Rarely will a specialist on disease causing ectoparasites, have the time to deal with phylogenies, cladistic analyses, or the construction of higher classifications. Generally, he is so inundated with specimens that need his

[1] Department of Health and Human Services, Public Health Service, National Institutes of Health, National Institute of Allergy and Infectious Diseases, c/o Department of Entomology, Museum Support Center, Smithsonian Institution, Washington, D.C. 20560, USA. Present address: Institute of Arthropodology and Parasitology, Georgia Southern University, Statesboro, Georgia 30460, USA

immediate attention that there is simply not enough time to work on these interesting problems.

This is definitely the case within the Ixodida, and it is a situation to be deplored. There are several problem areas in the systematics of the Ixodida that merit attention, and I shall present here one man's roster of those most deserving of study. To appreciate the scope of these problems, a listing of the valid taxonomic names of the Ixodida is a necessity. Although Kolonin (1978, 1981, 1983, 1984) published distribution lists for the Ixodidae, a complete taxonomic listing of all known tick species has not been attempted since Neumann (1911). My list of valid tick names is presented in Table 1. I realize that such a list is controversial and that probably no two tick taxonomists will ever agree on a single list of valid names. Nevertheless, Table 1 represents my current thinking on the matter and I welcome comments and suggestions to improve it.

Table 1. List of valid tick names

Nuttalliellidae

Nuttalliella
N. namaqua Bedford, 1931

Argasidae

Antricola

A. cernyi de la Cruz, 1978
A. coprophilus (McIntosh, 1935)
A. granasi de la Cruz, 1973
A. habanensis de la Cruz, 1976
A. marginatus (Banks, 1910)
A. martelorum de la Cruz, 1978
A. mexicanus Hoffman, 1959
A. naomiae de la Cruz, 1978
A. occidentalis de la Cruz, 1978
A. silvai Cerny, 1967

Argas

A. abdussalami Hoogstraal & McCarthy, 1965
A. africolumbae Hoogstraal, Kaiser, Walker, Ledger & Converse, 1975
A. arboreus Kaiser, Hoogstraal & Kohls, 1964
A. assimilis Teng & Song, 1983
A. australiensis Kohls & Hoogstraal, 1962
A. beijingensis Teng, 1983
A. beklemischevi Pospelova-Shtrom, Vasil'yeva & Semashko, 1963
A. boueti Roubaud & Colas-Belcour, 1933
A. brevipes Banks, 1908
A. brumpti Neumann, 1907
A. bureschi Dryenski, 1957
A. ceylonensis Hoogstraal & Kaiser, 1968
A. confusus Hoogstraal, 1955
A. cooleyi Kohls & Hoogstraal, 1960
A. cordiformis Hoogstraal & Kohls, 1967
A. cucumerinus Neumann, 1901
A. dalei Clifford, Keirans, Hoogstraal & Corwin, 1976
A. daviesi Kaiser & Hoogstraal, 1973
A. dewae Kaiser & Hoogstraal, 1974
A. dulus Keirans, Clifford & Capriles, 1971
A. echinops Hoogstraal, Uilenberg & Blanc, 1967
A. falco Kaiser & Hoogstraal, 1973
A. giganteus Kohls & Clifford, 1968
A. gilcolladoi Estrada-Peña, Lucientes & Sánchez, 1987
A. hermanni Audouin, 1827
A. himalayensis Hoogstraal & Kaiser, 1972
A. hoogstraali Morel & Vassiliades, 1965
A. japonicus Yamaguti, Clifford & Tipton, 1968
A. lagenoplastis Froggatt, 1906
A. latus Filippova, 1961
A. lowryae Kaiser & Hoogstraal, 1975
A. macrodermae Hoogstraal, Moorhouse, Wolf & Wassef, 1977
A. macrostigmatus Filippova, 1961
A. magnus Neumann, 1896
A. miniatus Koch, 1844
A. monachus Keirans, Radovsky & Clifford, 1973
A. moreli Keirans, Hoogstraal & Clifford, 1979
A. neghmei Kohls & Hoogstraal, 1961
A. nullarborensis Hoogstraal & Kaiser, 1973

Table 1. (Continued)

A. persicus (Oken, 1818)
A. polonicus Siuda, Hoogstraal, Clifford & Wassef, 1979
A. pusillus Kohls, 1950
A. radiatus Railliet, 1893
A. reflexus (Fabricius, 1794)
A. ricei Hoogstraal, Kaiser, Clifford & Keirans, 1975
A. robertsi Hoogstraal, Kaiser & Kohls, 1968
A. sanchezi Dugès, 1891
A. sinensis Jeu & Zhu, 1983
A. streptopelia Kaiser, Hoogstraal & Horner, 1970
A. striatus Bedford, 1932
A. theilerae Hoogstraal & Kaiser, 1970
A. transgariepinus (White, 1846)
A. transversus Banks, 1902
A. tridentatus Filippova, 1961
A. vespertilionis (Latreille, 1802)
A. vulgaris Filippova, 1961
A. walkerae Kaiser & Hoogstraal, 1969
A. zumpti Hoogstraal, Kaiser & Kohls, 1968

Nothoaspis

N. reddelli Keirans & Clifford, 1975

Ornithodoros

O. acinus Whittick, 1938
O. alactagalis Issakjan, 1936
O. amblus Chamberlain, 1920
O. apertus Walton, 1962
O. arenicolous Hoogstraal, 1953
O. asperus Warburton, 1918
O. azteci Matheson, 1935
O. batuensis Hirst, 1929
O. boliviensis Kohls & Clifford, 1964
O. braziliensis Aragão, 1923
O. brodyi Neumann, 1901
O. canestrinii (Birula, 1895)
O. capensis Neumann, 1901
O. casebeeri Jones & Clifford, 1972
O. chironectes Jones & Clifford, 1972
O. chiropterphila Dhanda & Rajagopalan, 1971
O. cholodkovskyi Pavlovsky, 1930
O. clarki Jones & Clifford, 1972
O. collocaliae Hoogstraal, Kadarsan, Kaiser & Van Peenan, 1974
O. compactus Walton, 1962
O. concanensis Cooley & Kohls, 1941
O. coniceps Canestrini, 1890
O. cooleyi McIvor, 1941
O. coriaceus Koch, 1844
O. cyclurae de la Cruz, 1984
O. darwini Kohls, Clifford & Hoogstraal, 1969
O. delanoei Roubaud & Colas-Belcour, 1931
O. denmarki Kohls, Sonenshine & Clifford, 1965
O. dugesi Mazzotti, 1943
O. dusbabeki Cerny, 1967
O. dyeri Cooley & Kohls, 1940
O. eboris Theiler, 1959
O. echimys Kohls, Clifford & Jones, 1969
O. elongatus Kohls, Clifford & Sonenshine, 1965
O. eptesicus Kohls, Clifford & Jones, 1969
O. eremicus Cooley & Kohls, 1941
O. erraticus (Lucas, 1846)
O. faini Hoogstraal, 1960
O. foleyi Parrot, 1928
O. furcosus Neumann, 1907
O. galapagensis Kohls, Clifford & Hoogstraal, 1969
O. graingeri Heisch & Guggisberg, 1953
O. grenieri Klein, 1965
O. gurneyi Warburton, 1926
O. hasei (Schulze, 1935)
O. hermsi Wheeler, Herms & Meyer, 1935
O. indica Rau & Rao, 1971
O. kelleyi Cooley & Kohls, 1941
O. knoxjonesi Jones & Clifford, 1972
O. lahorensis Neumann, 1908
O. macmillani Hoogstraal & Kohls, 1966
O. madagascariensis Hoogstraal, 1962
O. marinkellei Kohls, Clifford & Jones, 1969
O. maritimus Vermeil & Marguet, 1967
O. marmosae Jones & Clifford, 1972
O. marocanus Velu, 1919
O. mimon Kohls, Clifford & Jones, 1969
O. mormoops Kohls, Clifford & Jones, 1969
O. moubata (Murray, 1877)
O. muesebecki Hoogstraal, 1969
O. natterei Warburton, 1927
O. nicollei Mooser, 1932
O. normandi Larrousse, 1923
O. parkeri Cooley, 1936
O. peringueyi Bedford & Hewitt, 1925
O. peropteryx Kohls, Clifford & Jones, 1969
O. peruvianus Kohls, Clifford & Jones, 1969
O. peusi (Schulze, 1943)
O. piriformis Warburton, 1918
O. porcinus domesticus Walton, 1962
O. porcinus porcinus Walton, 1962
O. procaviae Theodor & Costa, 1960
O. puertoricensis Fox, 1947
O. rennellensis Clifford & Sonenshine, 1962
O. rossi Kohls, Sonenshine & Clifford, 1965

Table 1. (Continued)

O. rostratus Aragão, 1911
O. rudis Karsch, 1880
O. salahi Hoogstraal, 1953
O. savignyi (Audouin, 1826)
O. sawaii Kitaoka & Suzuki, 1973
O. solomonis Dumbleton, 1959
O. sonrai Sautet & Witkowski, 1943
O. sparnus Kohls & Clifford, 1963
O. spheniscus Hoogstraal, Wassef, Hays & Keirans, 1985
O. stageri Cooley & Kohls, 1941
O. steini (Schulze, 1935)
O. tadaridae Cerny & Dusbabek, 1967
O. talaje (Guèrin-Mènville, 1849)
O. tartakovskyi Olenev, 1931
O. tholozani (Laboulbene & Megnin, 1882)
O. tiptoni Jones & Clifford, 1972
O. turicata (Dugès, 1876)
O. tuttlei Jones & Clifford, 1972
O. vansomereni Keirans, Hoogstraal & Clifford, 1977
O. viguerasi Cooley & Kohls, 1941
O. yumatensis Cooley & Kohls, 1941
O. yunkeri Keirans, Clifford & Hoogstraal, 1984
O. zumpti Heisch & Guggisberg, 1953

Otobius

O. lagophilus Cooley & Kohls, 1940
O. megnini (Dugès, 1884)

Ixodidae

Amblyomma

A. albolimbatum Neumann, 1907
A. albopictum Neumann, 1899
A. americanum (Linnaeus, 1758)
A. antillorum Kohls, 1969
A. arianae Keirans & Garris, 1986
A. astrion Dönitz, 1909
A. auricularium (Conil, 1878)
A. australiense Neumann, 1905
A. babirussae Schulze, 1933
A. boulengeri Hirst & Hirst, 1910
A. brasiliense Aragão, 1908
A. cajennense (Fabricius, 1787)
A. calabyi Roberts, 1963
A. calcaratum Neumann, 1899
A. chabaudi Rageau, 1964
A. clypeolatum Neumann, 1899
A. coelebs Neumann, 1899
A. cohaerens Dönitz, 1909
A. compressum (Macalister, 1872)
A. cooperi Nuttall & Warburton, 1908
A. cordiferum Neumann, 1899
A. crassum Robinson, 1926
A. crenatum Neumann, 1899
A. cruciferum Neumann, 1901
A. cyprium Neumann, 1899
A. darwini Hirst & Hirst, 1910
A. dissimile Koch, 1844
A. eburneum Gerstäcker, 1873
A. echidnae Roberts, 1953
A. extraoculatum Neumann, 1899
A. falsomarmoreum Tonelli-Rondelli, 1933
A. fulvum Neumann, 1899
A. geayi Neumann, 1899
A. gemma Dönitz, 1909
A. geoemydae (Cantor, 1847)
A. goeldii (Neumann, 1899)
A. hainanense Teng, 1981
A. hebraeum Koch, 1844
A. helvolum Koch, 1844
A. humerale Koch, 1844
A. imitator Kohls, 1958
A. incisum Neumann, 1906
A. inornatum (Banks, 1909)
A. integrum Karsch, 1879
A. javanense (Supino, 1897)
A. laticaudae Warburton, 1933
A. lepidum Dönitz, 1909
A. limbatum Neumann, 1899
A. loculosum Neumann, 1907
A. longirostre (Koch, 1844)
A. macfarlandi Keirans, Hoogstraal & Clifford, 1973
A. macropi Roberts, 1953
A. maculatum Koch, 1844
A. marmoreum Koch, 1844
A. moreliae (Koch, 1867)
A. moyi Roberts, 1953
A. multipunctum Neumann, 1899
A. naponense (Packard, 1869)
A. neumanni Ribaga, 1902
A. nitidum Hirst & Hirst, 1910
A. nodosum Neumann, 1899
A. nuttalli Dönitz, 1909
A. oblongoguttatum Koch, 1844
A. ovale Koch, 1844
A. pacae Aragão, 1908
A. papuanum Hirst, 1914
A. parvitarsum Neumann, 1901
A. parvum Aragão, 1908
A. paulopunctatum Neumann, 1899
A. pecarium Dunn, 1933
A. personatum Neumann, 1901
A. pictum Neumann, 1906
A. pilosum Neumann, 1899
A. pomposum Dönitz, 1909

Table 1. (Continued)

A. postoculatum Neumann, 1899
A. pseudoconcolor Aragão, 1908
A. pseudoparvum Guglielmone, Mangold & Keirans, 1990
A. rhinocerotis (de Geer, 1778)
A. robinsoni Warburton, 1927
A. rotundatum Koch, 1844
A. sabanerae Stoll, 1890
A. scalpturatum Neumann, 1906
A. scutatum Neumann, 1899
A. sparsum Neumann, 1899
A. splendidum Giebel, 1877
A. squamosum Kohls, 1953
A. striatum Koch, 1844
A. supinoi Neumann, 1905
A. sylvaticum (de Geer, 1778)
A. tapirellum Dunn, 1933
A. testudinarium Koch, 1844
A. testudinis (Conil, 1877)
A. tholloni Neumann, 1899
A. tigrinum Koch, 1844
A. torrei Pèrez Viqueras, 1934
A. triguttatum Koch, 1844
A. triste Koch, 1844
A. tuberculatum Marx, 1894
A. usingeri Keirans, Hoogstraal & Clifford, 1973
A. variegatum (Fabricius, 1794)
A. varium Koch, 1844
A. williamsi Banks, 1924

Anocentor

Herein considered a subgenus of *Dermacentor* – see *D. nitens* Neumann, 1897

Anomalohimalaya

A. cricetuli Teng & Huang, 1981
A. lama Hoogstraal, Kaiser & Mitchell, 1970
A. lotozkyi Filippova & Panova, 1978

Aponomma

A. auruginans Schulze, 1936
A. colasbelcouri Santos Dias, 1958
A. concolor Neumann, 1899
A. crassipes Neumann, 1901
A. decorosum (Koch, 1844)
A. elaphense Price, 1959
A. exornatum (Koch, 1844)
A. fimbriatum (Koch, 1844)
A. flavamaculatum (Lucas, 1846)
A. fuscolineatum (Lucas, 1847)
A. gervaisi (Lucas, 1847)
A. hydrosauri (Denny, 1843)
A. komodoense Oudemans, 1929
A. kraneveldi Anastos, 1956
A. latum (Koch, 1844)
A. oudemansi Neumann, 1910
A. pattoni Neumann, 1910
A. quadricavum Schulze, 1941
A. soembawensis Anastos, 1956
A. sphenodonti Dumbleton, 1943
A. transversale (Lucas, 1845)
A. trimaculatum (Lucas, 1878)
A. varanensis (Supino, 1897)

Boophilus

B. annulatus (Say, 1821)
B. decoloratus (Koch, 1844)
B. geigyi Aeschlimann & Morel, 1965
B. kohlsi Hoogstraal & Kaiser, 1960
B. microplus (Canestrini, 1887)

Cosmiomma

C. hippopotamensis (Denny, 1843)

Dermacentor

D. albipictus (Packard, 1869)
D. andersoni Stiles, 1908
D. asper Arthur, 1960
D. atrosignatus Neumann, 1906
D. auratus Supino, 1897
D. circumguttatus Neumann, 1897
D. compactus Neumann, 1901
D. dispar Cooley, 1937
D. dissimilis Cooley, 1947
D. everestianus Hirst, 1926
D. halli McIntosh, 1931
D. hunteri Bishopp, 1912
D. imitans Warburton, 1933
D. latus Cooley, 1937
D. marginatus (Sulzer, 1776)
D. montanus Filippova & Panova, 1974
D. nitens Neumann, 1897
D. niveus Neumann, 1897
D. nuttalli Olenev, 1929
D. occidentalis Marx, 1892
D. parumapertus Neumann, 1901
D. pavlovskyi Olenev, 1927
D. pomerantzevi Serdyukova, 1951
D. raskemensis Pomerantsev, 1946
D. reticulatus (Fabricius, 1794)
D. rhinocerinus (Denny, 1843)
D. silvarum Olenev, 1931
D. sinicus Schulze, 1932
D. steini Schulze, 1932

Table 1. (Continued)

D. taiwanensis Sugimoto, 1935
D. ushakovae Filippova & Panova, 1987
D. variabilis (Say, 1821)

Haemaphysalis

H. aborensis Warburton, 1913
H. aciculifer Warburton, 1913
H. aculeata Lavarra, 1905
H. adleri Feldman-Muhsam, 1951
H. anomala Warburton, 1913
H. anoplos Hoogstraal, Uilenberg & Klein, 1967
H. aponommoides Warburton, 1913
H. asiatica (Supino, 1897)
H. atherurus Hoogstraal, Trapido & Kohls, 1965
H. bancrofti Nuttall & Warburton, 1915
H. bandicota Hoogstraal & Kohls, 1965
H. bartelsi Schulze, 1938
H. bequaerti Hoogstraal, 1956
H. birmaniae Supino, 1897
H. bispinosa Neumann, 1897
H. borneata Hoogstraal, 1971
H. bremneri Roberts, 1953
H. calcarata Neumann, 1902
H. calvus Nuttall & Warburton, 1915
H. campanulata Warburton, 1908
H. canestrinii (Supino, 1897)
H. capricornis Hoogstraal, 1966
H. caucasica Olenev, 1928
H. celebensis Hoogstraal, Trapido & Kohls, 1965
H. chordeilis (Packard, 1869)
H. concinna Koch, 1844
H. cooleyi Bedford, 1929
H. cornigera Neumann, 1897
H. cornupunctata Hoogstraal & Varma, 1962
H. cuspidata Warburton, 1910
H. danieli Cerny & Hoogstraal, 1977
H. darjeeling Hoogstraal & Dhanda, 1970
H. davisi Hoogstraal, Dhanda & Bhat, 1970
H. doenitzi Warburton, & Nuttall, 1909
H. elongata Neumann, 1897
H. erinacei Pavesi, 1884
H. eupleres Hoogstraal, Kohls & Trapido, 1965
H. flava Neumann, 1987
H. formosensis Neumann, 1913
H. fossae Hoogstraal, 1953
H. fujisana Kitaoka, 1970
H. garhwalensis Dhanda & Bhat, 1968
H. goral Hoogstraal, 1970
H. heinrichi Schulze, 1939
H. himalaya Hoogstraal, 1966
H. hirsuta Hoogstraal, Trapido & Kohls, 1966
H. hispanica Gil Collado, 1938
H. hoodi Warburton & Nuttall, 1909
H. hoogstraali Kohls, 1950
H. houyi Nuttall & Warburton, 1915
H. howletti Warburton, 1913
H. humerosa Warburton & Nuttall, 1909
H. hylobatis Schulze, 1933
H. hyracophila Hoogstraal, Walker & Neitz, 1971
H. hystricis Supino, 1897
H. ias Nakamura & Yajima, 1937
H. indica Warburton, 1910
H. indoflava Dhanda & Bhat, 1968
H. inermis Birula, 1895
H. intermedia Warburton & Nuttall, 1909
H. japonica Warburton, 1908
H. juxtakochi Cooley, 1946
H. kadarsani Hoogstraal & Wassef, 1977
H. kashmirensis Hoogstraal & Varma, 1962
H. kinneari Warburton, 1913
H. kitaokai Hoogstraal, 1969
H. koningsbergeri Warburton & Nuttall, 1909
H. kopetdaghica Kervabaev, 1962
H. kutchensis Hoogstraal & Trapido, 1963
H. kyasanurensis Trapido, Hoogstraal & Rajagopalan, 1964
H. lagostrophi Roberts, 1963
H. lagrangei Larrousse, 1925
H. leachi (Audouin, 1827)
H. lemuris Hoogstraal, 1953
H. leporispalustris (Packard, 1869)
H. longicornis Neumann, 1901
H. luzonensis Hoogstraal & Parrish, 1968
H. madagascariensis Colas-Belcour & Millot, 1948
H. mageshimaensis Saito & Hoogstraal, 1973
H. megalaimae Rajagopalan, 1963
H. megaspinosa Saito, 1969
H. minuta Kohls, 1950
H. mjoebergi Warburton, 1926
H. montgomeryi Nuttall, 1912
H. moreli Camicas, Hoogstraal & El Kammah, 1972
H. nadchatrami Hoogstraal, Trapido & Kohls, 1965
H. nepalensis Hoogstraal, 1962
H. nesomys Hoogstraal, Uilenberg & Klein, 1966
H. norvali Hoogstraal & Wassef, 1983
H. novaequineae Hirst, 1914
H. obesa Larrousse, 1925
H. obtusa Dönitz, 1910

Table 1. (Continued)

H. orientalis Nuttall & Warburton, 1915
H. ornithophila Hoogstraal & Kohls, 1959
H. palawanensis Kohls, 1950
H. papuana Thorell, 1882
H. paraleachi Camicas, Hoogstraal & El Kammah, 1983
H. paraturturis Hoogstraal, Trapido & Rebello, 1963
H. parmata Neumann, 1905
H. parva Neumann, 1908
H. pavlovskyi Pospelova-Shtrom, 1935
H. pedetes Hoogstraal, 1972
H. pentalagi Pospelova-Shtrom, 1935
H. petrogalis Roberts, 1970
H. phasiana Saito, Hoogstraal & Wassef, 1974
H. pospelovashtromae Hoogstraal, 1966
H. psalistos Hoogstraal, Kohls & Parrish, 1967
H. punctaleachi Camicas, Hoogstraal & El Kammah, 1973
H. punctata Canestrini & Fanzago, 1878
H. ramachandrai Dhanda, Hoogstraal & Bhat, 1970
H. ratti Kohls 1958
H. renschi Schulze, 1933
H. roubaudi Toumanoff, 1940
H. rugosa Santos Dias, 1956
H. rusae Kohls, 1950
H. sambar Hoogstraal, 1971
H. sciuri Kohls, 1950
H. semermis Neumann, 1901
H. shimoga Trapido & Hoogstraal, 1964
H. silacea Robinson, 1912
H. silvafelis Hoogstraal & Trapido, 1963
H. simplex Neumann, 1897
H. simplicima Hoogstraal & Wassef, 1979
H. spinigera Neumann, 1897
H. spinulosa Neumann, 1906
H. subelongata Hoogstraal, 1953
H. subterra Hoogstraal, El Kammah & Camicas, 1974
H. sulcata Canestrini & Fanzago, 1878
H. sumatraensis Hoogstraal, El Kammah, Kadarsan & Anastos, 1971
H. sundrai Sharif, 1928
H. susphilippensis Hoogstraal, Kohls & Parrish, 1968
H. taiwana Sugimoto, 1936
H. tauffliebi Morel, 1965
H. theilerae Hoogstraal, 1953
H. tibetensis Hoogstraal, 1965
H. tiptoni Hoogstraal, 1953
H. toxopei Warburton, 1927
H. traguli Oudemans, 1928
H. traubi Kohls, 1955
H. turturis Hoogstraal & Trapido, 1963
H. verticalis Itogaki, Noda & Yamaguchi, 1944
H. vidua Warburton & Nuttall, 1909
H. vietnamensis Hoogstraal & Wilson, 1966
H. warburtoni Nuttall, 1912
H. wellingtoni, Nuttall & Warburton, 1908
H. yeni Toumanoff, 1944
H. zumpti Hoogstraal & El Kammah, 1974

Hyalomma

H. aegyptium (Linnaeus, 1758)
H. albiparmatum Schulze, 1919
H. anatolicum anatolicum Koch, 1844
H. anatolicum excavatum Koch, 1844
H. arabica Pegram, Hoogstraal & Wassef, 1982
H. asiaticum Schulze & Schlottke, 1929
H. brevipunctata Sharif, 1928
H. detritum Schulze, 1919
H. dromedarii Koch, 1844
H. erythraeum Tonelli-Rondelli, 1932
H. franchinii Tonelli-Rondelli, 1932
H. hussaini Sharif, 1928
H. impeltatum Schulze & Schlottke, 1930
H. impressum Koch, 1844
H. lusitanicum Koch, 1844
H. kumari Sharif, 1928
H. marginatum marginatum Koch, 1844
H. marginatum isaaci Sharif, 1928
H. marginatum turanicum Pomerantzev, 1946
H. nitidum Schulze, 1919
H. punt Hoogstraal, Kaiser & Pedersen, 1969
H. rhipicephaloides Neumann, 1901
H. rufipes Koch, 1844
H. schulzei Olenev, 1931
H. truncatum Koch, 1844

Ixodes

I. acuminatus Neumann, 1901
I. acutitarsus (Karsch, 1899)
I. affinis Neumann, 1899
I. albignaci Uilenberg & Hoogstraal, 1969
I. alluaudi Neumann, 1913
I. amarali Fonesca, 1935
I. amersoni, Kohls, 1956
I. anatis Chilton, 1904
I. andinus Kohls, 1956
I. angustus Neumann, 1899
I. antechini Roberts, 1960
I. apronophorus Schulze, 1924

Table 1. (Continued)

I. arabukeiensis Arthur, 1959
I. arboricola Schulze & Schlottke, 1930
I. arebiensis Arthur, 1956
I. asanumai Kitaoka, 1973
I. aulacodi Arthur, 1956
I. auriculaelongae Arthur, 1958
I. auritulus Neumann, 1904
I. australiensis Neumann, 1904
I. baergi Cooley & Kohls, 1942
I. bakeri Arthur & Clifford, 1961
I. banksi Bishopp, 1911
I. bedfordi Arthur, 1959
I. bequaerti Cooley & Kohls, 1945
I. berlesei Birula, 1895
I. boliviensis Neumann, 1904
I. brewsterae Keirans, Clifford & Walker, 1982
I. browningi Arthur, 1956
I. brumpti Morel, 1965
I. brunneus Koch, 1844
I. calcarhebes Arthur & Zulu, 1980
I. caledonicus Nuttall, 1910
I. canisuga Johnston, 1849
I. capromydis Cerny, 1966
I. catherinei Keirans, Clifford & Walker, 1982
I. cavipalpus Nuttall & Warburton, 1908
I. ceylonensis Kohls, 1950
I. chilensis Kohls, 1957
I. colasbelcouri Arthur, 1957
I. collocaliae Schulze, 1937
I. conepati Cooley & Kohls, 1943
I. confusus Roberts, 1960
I. cookei Packard, 1869
I. cooleyi Aragão & Fonseca, 1951
I. copei Wilson, 1980
I. cordifer Neumann, 1908
I. cornuae Arthur, 1960
I. cornuatus Roberts, 1960
I. corwini Keirans, Clifford & Walker, 1982
I. crenulatus Koch, 1844
I. cuernavacensis Kohls & Clifford, 1966
I. cumulatimpunctatus Schulze, 1943
I. dammini Spielman, Clifford, Piesman & Corwin, 1979
I. dampfi Cooley, 1943
I. daveyi Nuttall, 1913
I. dawesi Arthur, 1956
I. dendrolagi Wilson, 1967
I. dentatus Marx, 1899
I. diomedeae Arthur, 1958
I. diversifossus Neumann, 1899
I. djaronensis Neumann, 1907
I. domerguei Uilenberg & Hoogstraal, 1965
I. downsi Kohls, 1957
I. drakensbergensis Clifford, Theiler & Baker, 1975
I. eadsi Kohls & Clifford, 1964
I. eastoni Keirans & Clifford, 1983
I. eichhorni Nuttall, 1916
I. eldaricus Djaparidze, 1950
I. elongatus Bedford, 1929
I. eudyptidis Maskell, 1885
I. euplecti Arthur, 1958
I. evansi Arthur, 1956
I. fecialis Warburton & Nuttall, 1909
I. festai Tonelli-Rondelli, 1926
I. fossulatus Neumann, 1899
I. frontalis Panzer, 1798
I. fuscipes Koch, 1844
I. galapagoensis Clifford & Hoogstraal, 1980
I. ghilarovi Filippova & Panova, 1988
I. gibbosus Nuttall, 1916
I. granulatus Supino, 1897
I. guatemalensis Kohls, 1957
I. hearlei Gregson, 1941
I. heinrichi Arthur, 1962
I. hexagonus Leach, 1815
I. himalayensis Dhanda & Kulkarni, 1969
I. hirsti Hassall, 1931
I. holocyclus Neumann, 1899
I. hoogstraali Arthur, 1955
I. howelli Cooley & Kohls, 1938
I. hyatti Clifford, Hoogstraal & Kohls, 1971
I. hydromyidis Swan, 1931
I. jacksoni Hoogstraal, 1967
I. jellisoni Cooley & Kohls, 1938
I. jonesae Kohls, Sonenshine & Clifford, 1969
I. kaiseri Arthur, 1957
I. kashimiricus Pomerantsev, 1948
I. kazakstani Olenev & Sorokoumov, 1934
I. kerguelenensis Andrè & Colas-Belcour, 1942
I. kingi Bishopp, 1911
I. kohlsi Aruthur, 1955
I. kopsteini Oudemans, 1926
I. kuntzi Hoogstraal & Kohls, 1965
I. laguri Olenev, 1931
I. lasallei Mendez Arocha & Ortiz, 1958
I. latus Arthur, 1958
I. laysanensis Wilson, 1964
I. lemuris Arthur, 1958
I. lewisi Arthur, 1965
I. lividus Koch, 1844
I. longiscutatum Boero, 1944
I. loricatus Neumann, 1899
I. loveridgei Arthur, 1958
I. luciae Senevet, 1940
I. lunatus Neumann, 1907

Table 1. (Continued)

I. luxuriosus Schulze, 1932
I. macfarlanei Keirans, Clifford & Walker, 1982
I. malayensis Kohls, 1962
I. marmotae Cooley & Kohls, 1938
I. marxi Banks, 1908
I. maslovi Emel'yanova & Kozlovskaya, 1967
I. matopi Spickett, Keirans, Norval & Clifford, 1981
I. mexicanus Cooley & Kohls, 1942
I. minor Neumann, 1902
I. minutae Arthur, 1959
I. mitchelli Kohls, Clifford & Hoogstraal, 1970
I. monospinosus Saito, 1968
I. montoyanus Cooley, 1944
I. moreli Arthur, 1957
I. moschiferi Nemenz, 1968
I. muniensis Arthur & Burrow, 1957
I. muris Bishopp & Smith, 1937
I. murreleti Cooley & Kohls, 1945
I. myospalacis Teng, 1986
I. myotomys Clifford & Hoogstraal, 1970
I. myrmecobii Roberts, 1962
I. nairobiensis Nuttall, 1916
I. nchisiensis Arthur, 1958
I. nectomys Kohls, 1957
I. neitzi Clifford, Walker & Keirans, 1977
I. neotomae Cooley, 1944
I. nesomys Uilenberg & Hoogstraal, 1969
I. neuguenensis Ringuelet, 1947
I. nipponensis Kitaoka & Saito, 1967
I. nitens Neumann, 1901
I. nuttalli Lahille, 1913
I. nuttallianus Schulze, 1930
I. occultus Pomerantsev, 1946
I. ochotonae Gregson, 1941
I. okapiae Arthur, 1956
I. oldi Nuttall, 1913
I. ornithorhynchi Lucas, 1846
I. ovatus Neumann, 1899
I. pacificus Cooley & Kohls, 1943
I. pararicinus Keirans & Clifford, 1985
I. pavlovskyi Pomerantsev, 1946
I. percavatus Neumann, 1906
I. peromysci Augustson, 1940
I. persulcatus Schulze, 1930
I. petauristae Warburton, 1933
I. philipi Keirans & Kohls, 1970
I. pilosus Koch, 1844
I. pomerantzevi Serdyukova, 1941
I. pomerantzi Kohls, 1957
I. priscicollaris Schulze, 1932
I. procaviae Arthur & Burrow, 1957
I. radfordi Kohls, 1948
I. rageaui Arthur, 1958
I. randrianasoloi Uilenberg & Hoogstraal 1969
I. rasus Neumann, 1899
I. redikorzevi Olenev, 1927
I. rhabdomysae Arthur, 1959
I. ricinus (Linnaeus, 1758)
I. rothschildi Nuttall & Warburton, 1911
I. rotundatus Arthur, 1958
I. rubicundus Neumann, 1904
I. rubidus Neumann, 1901
I. rugicollis Schulze & Schlottke, 1930
I. rugosus Bishopp, 1911
I. sachalinensis Filippova, 1971
I. scapularis Say, 1821
I. schillingsi Neumann, 1901
I. schulzei Aragão & Fonseca, 1951
I. sculptus Neumann, 1904
I. semenovi Olenev, 1929
I. shahi Clifford, Hoogstraal & Kohls, 1971
I. sigelos Keirans, Clifford & Corwin, 1976
I. signatus Birula, 1895
I. simplex Neumann, 1906
I. sinaloa Kohls & Clifford, 1966
I. sinensis Teng, 1977
I. soricis Gregson, 1942
I. spiculae Arthur, 1956
I. spinae Arthur, 1958
I. spinicoxalis Neumann, 1899
I. spinipalpis Hadwen & Nuttall, 1916
I. steini Schulze, 1932
I. stilesi Neumann, 1911
I. stromi Filippova, 1957
I. subterraneus Filippova, 1961
I. taglei Kohls, 1969
I. tamaulipas Kohls & Clifford,1966
I. tancitarius Cooley & Kohls, 1942
I. tanuki Saito, 1964
I. tapirus Kohls, 1957
I. tasmani Neumann, 1899
I. tecpanensis Kohls, 1957
I. texanus Banks, 1909
I. theilerae Arthur, 1958
I. thomasae Arthur & Burrow, 1957
I. tiptoni Kohls & Clifford, 1962
I. tovari Cooley, 1945
I. transvaalensis Clifford & Hoogstraal, 1966
I. trianguliceps Birula, 1895
I. trichosuri Roberts, 1960
I. tropicalis Kohls, 1957
I. turdus Nakatsuji, 1942
I. ugandanus Neumann, 1906
I. unicavatus Neumann, 1908
I. uriae White, 1852

Table 1. (Continued)

I. uruguayensis Kohls & Clifford, 1967
I. vanidicus Schulze, 1943
I. venezuelensis Kohls, 1945
I. ventalloi Gil Collado, 1936
I. vespertilionis Koch, 1844
I. vestitus Neumann, 1908
I. victoriensis Hirst, 1930
I. walkerae Clifford, Kohls & Hoogstraal, 1968
I. werneri Kohls, 1950
I. woodi Bishopp, 1911
I. zaglossi Kohls, 1960
I. zairensis Keirans, Clifford & Walker, 1982
I. zumpti Arthur, 1960

Margaropus

M. reidi Hoogstraal, 1956
M. wileyi Walker & Laurence, 1973
M. winthemi Karsch, 1879

Nosomma

N. monstrosum (Nuttall & Warburton, 1908)

Rhipicentor

R. bicornis Nuttall & Warburton, 1908
R. nuttalli Cooper & Robinson, 1908

Rhipicephalus

R. appendiculatus Neumann, 1901
R. arakeri Hiregoudar, 1975
R. armatus Pocock, 1900
R. arnoldi Theiler & Zumpt, 1949
R. bequaerti Zumpt, 1949
R. bergeoni Morel & Balis, 1976
R. boueti Morel, 1957
R. bursa canestrini & Fanzago, 1877
R. camelopardalis Walker & Wiley, 1959
R. camicasi Morel, Mouchet & Rodhain, 1976
R. capensis Koch, 1844
R. carnivoralis Walker, 1966
R. cliffordi Morel, 1964
R. complanatus Neumann, 1911
R. compositus Neumann, 1897
R. cuspidatus Neumann, 1906
R. deltoideus Neumann, 1910
R. distinctus Bedford, 1929
R. duttoni Neumann, 1907
R. dux Dönitz, 1910
R. evertsi evertsi Neumann, 1897
R. evertsi mimeticus Dönitz, 1910
R. follis Dönitz, 1910
R. fulvus Neumann, 1913
R. gertrudae Feldman-Muhsam, 1960
R. glabroscutatum du Toit, 1941
R. guilhoni Morel & Vassiliades, 1962
R. haemaphysaloides (Supino, 1897)
R. humeralis Tonelli-Rondelli, 1925
R. hurti Wilson, 1954
R. jeanneli Neumann, 1913
R. kochi Dönitz, 1905
R. leporis Pomerantsev, 1948
R. longiceps Warburton, 1912
R. longicoxatus Neumann, 1904
R. longus Neumann, 1907
R. lounsburyi Walker, 1990
R. lunulatus Neumann, 1907
R. maculatus Neumann, 1901
R. masseyi Nuttall & Warburton, 1908
R. moucheti Morel, 1964
R. muehlensi Zumpt, 1943
R. muhsamae Morel & Vassiliades, 1964
R. neumanni Walker, 1990
R. nitens Neumann, 1904
R. oculatus Neumann, 1901
R. pilans Schulze, 1935
R. planus Neumann, 1910
R. praetextatus Gerstäcker, 1873
R. pravus Dönitz, 1910
R. pulchellus Gerstäcker, 1873
R. pumilio Schulze, 1935
R. punctatus Warburton, 1912
R. pusillus Gil Collado, 1938
R. ramachandrai Dhanda, 1966
R. reichenowi Zumpt, 1943
R. rossicus Yakimov & Kol'-Yakimova, 1911
R. sanguineus (Latreille, 1806)
R. scalpturatus Santos Dias, 1959
R. schulzei Olenev, 1929
R. sculptus Warburton, 1912
R. senegalensis Koch, 1844
R. serrranoi Santos Dias, 1950
R. simpsoni Nuttall, 1910
R. simus Koch, 1844
R. sulcatus Neumann, 1908
R. supertritus Neumann, 1907
R. theileri Bedford & Hewitt, 1925
R. tricuspis, Donitz, 1906
R. turanicus Pomerantsev, 1936
R. zambeziensis Walker, Norval & Corwin, 1981
R. ziemanni Neumann, 1904
R. zumpti Santos Diaz, 1951

It should be emphasized that this listing is one man's opinion. It should also be remembered, as the the late professor George Wharton once said, "Species change from place to place and names change from time to time."

2 Nuttalliellidae Schulze

The monotypic family Nuttalliellidae is represented by *Nuttalliella namaqua* Bedford, known from Namibia, Republic of South Africa and Tanzania. It may not be a true link between the Argasidae and Ixodidae, but it would be fascinating if a colony of this rare species could be established so that we could see what the male, nymph and larva look like. The taxonomy of the female was reviewed by Keirans et al. (1976).

3 Argasidae Canestrini

There is debate about the number of genera within the Argasidae. I recognize five genera and shall not go into the theoretical aspects of the higher classification of argasids in this paper. For those interested in this subject, the publications of Clifford et al. (1964), Pospelova-Shtrom (1969), and Camicas and Morel (1977) are a good starting point.

Until the late 1960s the Neotropical genus *Antricola* Cooley and Kohls was composed of four species. Between 1973 and 1978 de la Cruz described six additional species from Cuba. These six species may or may not be valid. We have not been able to examine the type specimens, and if one looks at the illustrations (e.g. Fig. 1 of de la Cruz 1973), it is impossible to judge the validity of the species. This is an interesting group of bat ticks in which only the subadult stages are parasitic, but there is no published up to date review of the genus.

The genus *Argas* Latreille is currently composed of seven subgenera: *Argas*, *Carios*, *Chiropterargas*, *Microargas*, *Ogadenus*, *Persicargas*, and *Secretargas*. Birds are hosts for species of the subgenera *Argas* and *Persicargas*; cave-dwelling bats are parasitized by the species of *Carios* and *Chiropterargas*; the single species of *Microargas* is found on the Galapagos giant tortoise; mammals are the host for the single species of *Ogadenus*; and the three species of *Secretargas* are found on bats, lizards and insectivores.

The status of the subgenus *Argas* has been solidified by a series of publications by Hoogstraal and Kohls (1960a, b); Kohls and Hoogstraal (1960, 1961); Kohls et al. (1961); Filippova (1966); Clifford et al. (1978, 1979); and Keirans et al. (1979). New world *Persicargas* were reviewed by Kohls et al. (1970); Afrotropical species by Kaiser et al. (1964) and Kaiser and Hoogstraal (1969); and Oriental and Australian species by Hoogstraal et al. (1968) and Hoogstraal and Kaiser (1973). The subgenera *Carios*, *Chiropterargas* and *Secretargas* have not been summarized. For the single species within the subgenus *Ogadenus* see Hoogstraal (1956), and for the single species in the subgenus *Microargas* see Hoogstraal and Kohls (1966) and Hoogstraal et al. (1973). For the identification of larval *Argas* species see Sonenshine et al. (1962).

The monotypic genus *Nothoaspis* was described by Keirans and Clifford (1975) from a bat cave in Campeche, Mexico. Subsequently, nymphs and associated larvae were described from bat caves in Tabasco and Yucatan, Mexico (Keirans et al. 1977). To date that is the sum of our knowledge of this rare argasid genus.

The genus *Ornithodoros* Koch is the largest within the Argasidae and consists of about 100 species (Table 1). Divergent views on the subgeneric status of the genus were summarized by Clifford et al. (1964), Pospelova-Shtrom (1969), and Camicas and Morel (1977).

Within the genus *Ornithodoros* the larval stages are better known than the nymphs or adults because larvae of most *Ornithodoros* species feed for a long period and are therefore the stage most often collected. Many of the bat-feeding *Ornithodoros* are known only from the larval stage. Larval keys to western hemisphere species were given by Kohls et al. (1965) with a revised key including several new species by Jones and Clifford (1972). For a key to eastern hemisphere larvae, see Sonenshine et al. (1966). Filippova (1961, 1966) and Kusov (1973) provided an excellent overview of the *Ornithodoros* fauna of the USSR. No comprehensive treatment of the genus is available.

The genus *Otobius* Banks contains two species: *O. megnini*, a parasite of large wandering mammals in western North America that has also been introduced into South America, South Africa, India and several other areas of the world; and *O. lagophilus*, which feeds on the genera *Lepus* and *Sylvilagus* in the Great Basin area of western North America. Adults of this genus have nonfunctional mouthparts and are not parasitic. Keys separating the species can be found in Cooley and Kohls (1944).

4 Ixodidae Koch

The family Ixodidae is usually considered to be composed of approximately 13 genera in 5 subfamilies arranged as follows: Ixodinae – *Ixodes*; Amblyomminae – *Amblyomma, Aponomma*; Haemaphysalinae – *Haemaphysalis*; Hyalomminae – *Hyalomma*; and Rhipicephalinae – *Anomalohimalaya, Boophilus, Cosmiomma, Dermacentor, Margaropus, Nosomma, Rhipicentor*, and *Rhipicephalus*.

The genus *Anocentor* may or may not be valid; there are strong arguments for and against its validity. For the purposes of this chapter, I shall consider *Anocentor* to be a subgenus of *Dermacentor*.

There are approximately one hundred species in the genus *Amblyomma* Koch, and all appear to have a three-host life cycle. They inhabit tropical or subtropical areas in both hemispheres. A few species reach the Nearctic, but only three species are found in the Palearctic (Japan): *A. nitidum, A. geoemydae* and *A. testudinarium*.

The monographic treatment of the genus *Amblyomma* by Robinson (1926), although outdated and in need of revision, is still the finest coverage of the group ever published. More recent regional works by Matthysse and Colbo (1987) for the Afrotropical region, Jones et al. (1972) for the Neotropics and Nearctic, and Roberts (1970) for Australia, cover most areas where *Amblyomma* species are found. A good generic review for southeastern Asia, including the island fauna, is still needed.

Worldwide, adults of *Amblyomma* species are quite well known. However, especially in South America (where the largest number of species is found) knowledge of the immature stages is woefully inadequate and will only be acquired through intensive rearing studies.

Within the recently described genus *Anomalohimalaya* Hoogstraal, Kaiser and Mitchell, there are thus far, three described species: *A. lama* Hoogstraal et al. (1970), found on rodents in Nepal; *A. lotozkyi* Filippova and Panova (1978), a parasite of *Alticola argentatus* in Tadzhikstan, USSR; and *A. cricetuli* Teng and Huang (1981), a parasite of *Cricetulus migratorius* in the Xinjiang Uighur Autonomous Region, People's Republic of China.

Members of the genus *Aponomma* Neumann are similar to *Amblyomma* but lack eyes. These small ornamented (or sometimes unornamented) ticks are almost all parasites of snakes or varanid lizards (two species are found on marsupials in Australia). Adults and immatures of the approximately twenty-three species are often found on the same host.

Since the genus *Aponomma* was established in 1899, as many as 62 species names have been included in the genus. In a monographic revision of the genus, Kaufmann (1972) reduced this number to 21. A differing view on the arrangement of species within *Aponomma* was given by Santos Dias (1985). The only species found in the western hemisphere are *Aponomma quadricavum* and *A. elaphense* (Anderson et al. 1981; Keirans and Degenhardt 1985). All other species are African, Asian or Australian in origin.

It is almost universally agreed that there are five valid species of the genus *Boophilus* Curtice. These are economically important, one-host parasites of cattle, sheep, goats and other domesticated animals. The genus occurs worldwide but primarily south of the northern 40th parallel. The most important and widespread species is *B. microplus*, followed by *B. annulatus*, *B. decoloratus*, *B. geigyi* and *B. kohlsi*, the last named being a parasite of goats and sheep in Jordan and Israel. The four other species are all chiefly cattle parasites, *Boophilus geigyi* being found in West Africa, *B. decoloratus* in South and East Africa, and the remaining two species widespread. Nunez et al. (1985) presented a monographic treatment of *Boophilus microplus*, and Feldman-Muhsam and Shechter (1970) reviewed all species and provided distribution maps and keys to the adults.

The monotypic genus *Cosmiomma* Schulze, represented by *C. hippopotamensis* Denny, is a rare species found in East and South Africa. The preferred host appears to be the black rhinoceros, *Diceros bicornis*, and not the hippopotamus as one might guess from the species epithet. For a summary of the known hosts and distribution of this unusual and brightly ornamented species see Bezuidenhout and Schneider (1972).

The genus *Dermacentor* Koch is represented by about thrity-two species. There are 12 species in the new world, 2 species in Africa, and the remainder are Eurasian in origin.

The monographic treatment of this genus by Arthur (1960) is a seriously flawed work and, if consulted at all by tick workers, should be used with great caution. Cooley (1938) provided keys and illustrations to the United States species and Yunker et al. (1986) gave keys and scanning electron microscope photos for adults of all 12 New World species. For a key and illustrations to the two African species see Hoogstraal (1956). For keys and illustrations to the

Eurasian species including immatures see Filippova and Panova (1984, 1988, 1989) and Filippova et al. (1981, 1986). Berdyev (1989) presented an interesting theory on the origin and distribution of the genus *Dermacentor*. Teng (1982) reviewed the distribution of this genus in China. In a series of ten papers between 1983 and 1988 in the Journal of Medical Entomology, Hoogstraal and co-workers revised the *Dermacentor* species of Southeast Asia (Hoogstraal and Wassef 1984, 1985a, b; Wassef and Hoogstraal 1983, 1984a, b, 1986a, b, 1988; Hoogstraal et al. 1986).

The second largest tick genus is *Haemaphysalis* Koch with about 147 species, and all appear to be three-host parasites of birds and mammals.

During his long and distinguished career, Harry Hoogstraal was recognized as the undisputed authority on *Haemaphysalis*; however, he never produced a monographic revision or keys to the species, and both are badly needed. Trapido et al. (1964) and Tanskul and Inlao (1989) published keys to the species of South India and Thailand, respectively. The three New World species are well known (Cooley 1946), and Matthysse and Colbo (1987) provided a key to most of the African species. Because most of the taxonomic literature on *Haemaphysalis* is scattered, it would be wise to consult the list of publications by Hoogstraal (Keirans 1987) for guidance to the species. Using Hoogstraal's *Bibliography of Ticks and Tickborne Diseases*, Estrada-Pena (1989) published an index catalog to the *Haemaphysalis* of the world. Although replete with errors and uncritical acceptance of some publications, it can be of assistance in tracking down literature on the genus.

There are about 30 species of *Hyalomma* Koch found from the Mediterranean basin and Africa to the Indian subcontinent. These large, tough ticks are common in low altitude areas with long, hot, dry seasons.

Until the middle of this century *Hyalomma* taxonomy was in a chaotic state. For example, Delpy (1936), who spent a good deal of time studying the genus, came to the conclusion that diagnosis of species was almost impossible. Because of their great variability and despite several notable attempts to quantify and delimit species, identification of individual specimens can still be a daunting task.

Keys and illustrations to the *Hyalomma* species of the Indian subcontinent and neighboring regions were published by Kaiser and Hoogstraal (1963, 1964) and Singh and Dhanda (1965). Other regional works are Hoogstraal et al. (1981) for Saudi Arabia and adjacent areas; Hoogstraal and Kaiser (1959) (and others in the same series) for North Africa; and Hoogstraal (1956) for equatorial Africa. It should be noted that in the 1956 publication, Hoogstraal's discussion of *Hyalomma excavatum* (pp. 436–451) and illustration (Plate XIL) actually referred to *Hyalomma anatolicum anatolicum*.

The largest genus of ticks, comprising about 250 species, is *Ixodes* Latreille. All known species of *Ixodes* are three-host ticks and all except one are parasites of birds or mammals during at least part of their life cycle. The genus is found in climates that range from arctic to tropical.

Within what is considered herein the genus *Ixodes*, Camicas and Morel (1977) recognized two subfamilies: Eschatocephalinae containing the genera *Ceratixodes*, *Lepidixodes*, *Eschatocephalus*, and *Scaphixodes*; and Ixodinae containing the genus *Ixodes*. Clifford et al. (1973) recognized the single subfamily Ixodinae containing the genus *Ixodes*, and relegated other entities to the subgeneric level.

The first comprehensive revision of *Ixodes* was by Nuttall and Warburton (1911) and no revision of the genus has appeared since that time. Cooley and Kohls (1945) reviewed the North American species and Keirans and Clifford (1978) published a scanning electron micrograph key to the species found in the United States. There is no review available for the genus *Ixodes* in Central or South America, although at least forty-five species are known to occur there. A publication on the *Ixodes* of South America would be very welcome since the genus is so well represented in the Neotropics but so poorly known. Arthur (1965) summarized the *Ixodes* of Africa; for African *Ixodes* species since that time see Matthysse and Colbo (1987). For a discussion of the European representatives of the genus see Arthur (1963) and Babos (1964). Filippova (1977) presented an excellent overview of the *Ixodes* found in the USSR, as did Yamaguti et al. (1971) for Japan and Korea. For a review of the unique Australian *Ixodes* fauna, see Roberts (1970).

There are three species of the genus *Margaropus* Karsch: *M. reidi* and *M. wileyi* on giraffes in African lowland savanna areas, and *M. winthemi*, a rare species but one that can be found in large numbers at high altitudes in South Africa on the Cape mountain zebra, *Equus zebra zebra. Margaropus winthemi* is active in the winter and will also parasitize domestic cattle and horses. For a redescription of *M. winthemi* and *M. reidi* see Arthur (1960). For a description of *M. wileyi* and a key to adults of the three *Margaropus* species see Walker and Laurence (1973).

The genus *Nosomma* Schulze was proposed to accomodate the monotypic species *Nosomma monstrosum* (Nuttall and Warburton). This is a three-host tick with a 2-year life cycle; it is found on buffalo, horses, domestic cattle, boar, and occasionally man. Most collections of *N. monstrosum* come from India but it has also been found in Sri Lanka (Seneviratana 1965), Bangladesh (Nuttall and Warburton 1908), Laos (Toumanoff 1944) and Nepal (Hoogstraal 1970). Arthur and Chaudhuri (1965) redescribed and illustrated the adults of *N. monstrosum*, and the nymph and larva were described by Singh (1968).

The genus *Rhipicentor* Nuttall and Warburton is composed of two species, *R. bicornis* and *R. nuttalli*. The generic name *Rhipicentor* was used by Nuttall and Warburton because ticks of this genus combine morphological features of both *Rhipicephalus* and *Dermacentor*. Both species are found only in Africa south of the Sahara; *R. bicornis* in southern and central Africa on goats, cattle, horses, dogs, and wild carnivores, and *R. nuttalli* in southern Africa primarily on carnivores. For a review of this genus, descriptions of the larvae and a key to the adults see Theiler (1961).

Of the approximately 73 species of the genus *Rhipicephalus* Koch, the vast majority (ca. 54) are African, the remainder being found in Europe or Asia. One species, *R. sanguineus*, although African in origin now occurs around the globe from about 45° N to 45° S and is the most widespread tick in the world. Most rhipicephalids are three-host ticks but two-host species exist. These ticks prefer mammals of all sizes for their blood meals; birds are seldom utilized as hosts.

The literature on the genus *Rhipicephalus* is widely scattered and no comprehensive monograph is available to researchers. Matthysse and Colbo (1987) provided a key to the adults of thirty-eight species found in Uganda and neighboring areas. They realized that adult rhipicephalids vary in morphological characters and any investigator would be wise to carefully read their introductory paragraphs before attempting to use the keys. Others who have investigated the genus *Rhipicephalus* in Africa in an attempt to clarify the vexing species problems, include Zumpt (1949), Theiler and Robinson (1953), Tendeiro (1959), Walker (1961, 1974), Clifford and Anastos (1962, 1964), Yeoman and Walker (1967), Morel (1976), Clifford et al. (1983), Pegram et al. (1987a, b, c), and Walker et al. (1988). Although only in the Preliminary stages, J.B. Walker and J.E. Keirans plan to produce a handbook on the *Rhipicephalus* species of Africa, containing descriptions as well as pen and ink drawings and scanning electron photomicrographs for all stages of the known species. For Eurasian *Rhipicephalus* species, see Filippova (1981) and Filippova and Panova (1983).

References

Anderson JF, Magnarelli LA, Keirans JE (1981) *Aponomma quadricavum* (Acari: Ixodidae) collected from an imported boa, *Epicrates striatus*, in Connecticut. J Med Entomol 18:123–125

Arthur DR (1960) Ticks: a monograph of the Ixodoidea. Part V. On the genera *Dermacentor*, *Anocentor*, *Cosmiomma*, *Boophilus* and *Margaropus*. Cambridge University Press, Cambridge

Arthur DR (1963) British ticks. Butterworths, London, 213 pp

Arthur DR (1965) Ticks of the genus *Ixodes* in Africa. Athlone Press, London 348 pp

Arthur DR, Chaudhuri RP (1965) A revision of *Nosomma monstrosum* (Nuttall and Warburton, 1908) Ixodoidea: Ixodidae. Parasitology 55:391–400

Babos S (1964) Die Zeckenfauna Mitteleuropas. Akademiai Kiado, Budapest, 410 pp

Berdyev A (1989) The zoogeography and distribution of the genus *Dermacentor* Koch, 1844 (Parasitiformes, Ixodidae). Parazitologia 23:166–172 (in Russian).

Bezuidenhout JD, Schneider HP (1972) Studies on the biology of *Cosmiomma hippopotamensis* Denny, 1843 in South West Africa. JS Afr Vet Med Assoc 43:301–304

Camicas JL, Morel PC (1977) Position systematique et classification des tiques (Acarida: Ixodida). Acarologia 18:410–420

Clifford CM, Anastos G (1962) Ticks. Explor Parc Natl Upemba Miss GF de Witte (1946–49) 66:1–43

Clifford CM, Anastos G (1964) Ticks. Parc Natl Garamba Miss H de Saeger (1949–52) 44:1–40

Clifford CM, Kohls GM, Sonenshine DE (1964) The systematics of the subfamily Ornithodorinae (Acarina: Argasidae). I. The genera and subgenera. Ann Entomol Soc Am 57:429–437

Clifford CM, Sonenshine DE, Keirans JE, Kohls GM (1973) Systematics of the subfamily Ixodinae (Acarina: Ixodidae). I. The subgenera of *Ixodes*. Ann Entomol Soc Am 66:489–500

Clifford CM, Hoogstraal H, Keirans JE, Rice RCA, Dale W (1978) Observations on the subgenus *Argas* (Ixodoidea: Argasidae: *Argas*) 14. Identity and biological observations of *Argas* (*A.*) *cucumerinus* from Peruvian seaside cliffs and a summary of the status of the subgenus in the Neotropical faunal region. Ann Entomol Soc Am 15:57–73

Clifford CM, Keirans JE, Hoogstraal H, Morel PC (1979) Observations on the subgenus *Argas* (Ixodoidea: Argasidae: *Argas*) 15. Identity of *Argas* (*A.*) *magnus* from Ecuador and Colombia. Ann Entomol Soc Am 15:157–165

Clifford CM, Walker JB, Keirans JE (1983) Clarification of the status of *Rhipicephalus kochi* Donitz, 1905 (Ixodoidea, Ixodidae). Onderstepoort J Vet Res 50:77–89

Cooley RA (1938) The genera *Dermacentor* and *Otocentor* (Ixodidae) in the United States. Natl Inst Health Bull 171: 1–89

Cooley RA (1946) The genera *Boophilus*, *Rhipicephalus*, and *Haemaphysalis* (Ixodidae) of the New World. Natl Inst Health Bull 187:1–54
Cooley RA, Kohls GM (1944) The Argasidae of North America, Central America and Cuba. Am Midland Natl Monogr No 1. University Press, Notre Dame, Indiana 152 pp
Cooley RA, Kohls GM (1945) The genus *Ixodes* in North America. Natl Inst Health Bull 184:1–246
de la Cruz J (1973) Notas sobre las garrapatas del genero *Antricola* Cooley y Kohls, 1942 (Ixodiformes, Argasidae) con la descripcion de una nueva especie. Simp 30 Aniv Soc Esp Cuba Es Carsol (44):3–13 (in Spanish)
Delpy L (1936) Notes sur les ixodides du genre *Hyalomma* (Koch). Ann Parasitol Hum Comp 14:206–245 (in French)
Ehrlich PR (1961) Systematics in 1970: some unpopular predictions. Syst Zool 10:157–158
Estrada-Pena A (1989) Indece-catalogo de las garrapatas (Acarina: Ixodoidea) en el mundo. Vol 1. Genero *Haemaphysalis*. Universidad de Zaragoza, 932 pp (in Spanish)
Feldman-Muhsam B, Shechter R (1970) Some notes on the genus *Boophilus* (Ixodidae) with special reference to species found in Israel. J Med Entomol 7:677–686
Filippova NA (1961) Larvae and nymphs of ticks of the subfamily Ornithodorinae (Ixodoidea, Argasidae in the fauna of the Soviet Union. Parazitol Sborn Zool Inst Akad Nauk SSSR 20:148–182
Filippova NA (1966) Argasid ticks (Argasidae). Fauna SSSR, Paukoobraznye 4, 255 pp (in Russian)
Filippova NA (1977) Ixodid ticks of the subfamily Ixodinae. Fauna SSSR, Paukoobraznye 4, 396 pp (in Russian)
Filippova NA (1981) On the diagnosis of species of the genus *Rhipicephalus* Koch (Ixodoidea, Ixodidae) from the fauna of the USSR and adjoining territories by nymphal instar. Parazitol Sborn Zool Inst Akad Nauk SSSR 30:47–68 (in Russian with English summary)
Filippova NA, Panova IV (1978) *Anomalohimalaya lotozkyi* sp.n., a new species of ixodid tick from Peter I Ridge (Ixodoidea, Ixodidae). Parazitol Leningrad 12:391–339 (in Russian with English summary)
Filippova NA, Panova IV (1983) Importance of external ultrastructures for systematics of ixodid ticks of the genus *Rhipicephalus* Koch (Ixodoidea, Ixodidae). Parazitol Sborn Zool Inst Akad Nauk SSSR 31:17–36 (in Russian)
Filippova NA, Panova IV (1984) Contribution to diagnosis of subgenera of the genus *Dermacentor* Koch from larvae and nymphs and new data on distribution of the subgenus *Asiacentor* (Ixodidae). Parazitol Leningrad 18:135–139 (in Russian with English summary)
Filippova NA, Panova IV (1988) Description of larvae and nymphs of the ixodid tick *Dermacentor ushakovae*. Parazitol Leningrad 22:122–131 (in Russian with English summary)
Filippova NA, Panova IV (1989) Revision of the genus *Dermacentor* Koch of the fauna of the USSR and contiguous territories (Ixodoidea, Ixodidae). Parazitol Sborn Zool Inst Akad Nauk SSSR 35:49–95 (in Russian with English summary)
Filippova NA, Panova IV, Grebenjuk RV (1981) On the diagnosis of *Dermacentor* species from central Asia by the nymphal instar (Ixodidae). Parazitol Leningrad 15:441–450 (in Russian with English summary)
Filippova NA, Panova IV, Grebenjuk RV (1986) Diagnosis of species of the genus *Dermacentor* Koch from central Asia by the larval instar (Ixodidae). Parazitol Leningrad 22:89–98 (in Russian with English summary)
Hoogstraal H (1956) African Ixodoidea. I. Ticks of the Sudan (with special reference to Equatoria Province and with preliminary reviews of the genera *Boophilus*, *Margaropus*, and *Hyalomma*). Dept Navy, Bur Med Surg Washington, DC, USA, 1101 pp
Hoogstraal H (1970) Human infestation by ticks (Ixodidae) in the Himalaya. In: Singh KS, Tandon BK (eds) Srivastava commemoration volume. Div Parasitol Indian Vet Res Inst, Izatnagar, India, 75–89 pp
Hoogstraal H, Kaiser MN (1959) Observations on Egyptian *Hyalomma* ticks (Ixodoidea, Ixodidae). 5. Biological notes and differences in identity of *H. anatolicum* and its subspecies *anatolicum* Koch and *excavatum* Koch, among Russian and other workers. Identity of *H. luscitanicum* Koch. Ann Entomol Soc Am 52:243–261
Hoogstraal H, Kaiser MN (1973) The subgenus *Persicargas* (Ixodoidea: Argasidae: *Argas*). 18. *A.* (*P.*) *nullaborensis*, new species, from Western Australia. Ann Entomol Soc Am 66:1296–1298

Hoogstraal H, Kohls GM (1960a) Observations on the subgenus *Argas* (Ixodoidea, Argasidae, *Argas*). 1. Study of *A. reflexus reflexus* (Fabricius 1974), the European bird argasid. Ann Entomol Soc Am 53:611–618

Hoogstraal H, Kohls GM (1960b) Observations on the subgenus *Argas* (Ixodoidea, Argasidae, *Argas*). 3. A biological study of *A. reflexus hermanni* Audouin 1827 (revalidated), the African bird argasid. Ann Entomol Soc Am 53:743–755

Hoogstraal H, Kohls GM (1966) *Argas (Microargas) transversus* Banks (new subgenus) (Ixodoidea, Argasidae), a diminutive parasite of the Galapagos giant tortoise: Redescription of the holotype male and description of the larva. Ann Entomol Soc Am 59:247–252

Hoogstraal H, Wassef HY (1984) *Dermacentor (Indocentor) compactus* (Acari: Ixodoidea: Ixodidae): wild pigs and other hosts and distribution in Malaysia, Indonesia, and Borneo. J Med Entomol 21:174–178

Hoogstraal H, Wassef HY (1985a) *Dermacentor (Indocentor) auratus* (Acari: Ixodoidea: Ixodidae): hosts, distribution, and medical importance in tropical Asia. J Med Entomol 22:170–177

Hoogstraal H, Wassef HY (1985b) *Dermacentor (Indocentor) atrosignatus*) Acari: Ixodoidea: Ixodidae): hosts and distribution in the Malay Peninsula, Indonesia, Borneo, and southern Philippines. J Med Entomol 22:644–674

Hoogstraal H, Kaiser MN, Kohls GM (1968) The subgenus *Persicargas* (Ixodoidea, Argasidae, *Argas*). 4. *Argas (P.) robertsi*, new species, a parasite of Australian fowl, and keys to Australian argasid species. Ann Entomol Soc Am 61:535–539

Hoogstraal H, Kaiser MN, Mitchell RM (1970) *Anomalohimalaya lama*, new species (Ixodoidea, Ixodidae), a tick parasitizing rodents, shrews, and hares in the Tibetan highlands of Nepal. Ann Entomol Soc Am 63:1576–1585

Hoogstraal H, Clifford CM, Keirans JE (1973) *Argas (Microargas) transversus* (Ixodoidea: Argasidae) of Galapagos giant tortoises: description of the female and nymph. Ann Entomol Soc Am 66:727–732

Hoogstraal H, Wassef HY, Buttiker W (1981) Ticks (Acarina) of Saudi Arabia. Fam. Argasidae, Ixodidae. Fauna Saudi Arabia 3:25–110

Hoogstraal H, Wassef HY, Santana FJ, Kuntz RE (1986) *Dermacentor (Indocentor) taiwanensis* (Acari: Ixodoidea: Ixodidae): hosts and distribution in Taiwan and southern Japan. J Med Entomol 23:286–288

Jones EK, Clifford CM (1972) The systematics of the subfamily Ornithodorinae (Acarina: Argasidae). V. A revised key to larval Argasidae of the western hemisphere and descriptions of seven new species of *Ornithodoros*. Ann Entomol Soc Am 65:730–740

Jones EK, Clifford CM, Keirans JE, Kohls GM (1972) The ticks of Venezuela (Acarina: Ixodoidea) with a key to the species of *Amblyomma* in the western hemisphere. Brigham Young Univ Sci Bull Biol Ser 17:1–40

Kaiser MN, Hoogstraal H (1963) The *Hyalomma* ticks (Ixodoidea, Ixodidae) of Afghanistan. J Parasitol 49:130–139

Kaiser MN, Hoogstraal H (1964) The *Hyalomma* ticks (Ixodoidea, Ixodidae) of Pakistan, India and Ceylon, with keys to subgenera and species. Acarologia 6:257–286

Kaiser MN, Hoogstraal H (1969) The subgenus *Persicargas* (Ixodoidea, Argasidae, *Argas*). 7. *A.* (*P.*) *walkerae*, new species, a parasite of domestic fowl in southern Africa. Ann Entomol Soc Am 62:885–890

Kaiser MN, Hoogstraal H, Kohls GM (1964) The subgenus *Persicargas*, new subgenus (Ixodoidea, Argasidae, *Argas*. I. *A.* (*P.*) *arboreus*, new species, an Egyptian *persicus*-like parasite of wild birds, with a redefinition of the subgenus *Argas*. Ann Entomol Soc Am 57:60–69

Kaufmann TS (1972) A revision of the genus *Aponomma* Neumann, 1899 (Acarina: Ixodidae). PhD Thesis, University of Maryland, College Park, MD, USA 389 pp

Keirans JE (1987) Harry Hoogstraal (1917–1986): A bibliography. J Med Entomol 24:121–140

Keirans JE, Clifford CM (1975) *Nothoaspis reddelli*, new genus and new species (Ixodoidea: Argasidae), from a bat cave in Mexico. Ann Entomol Soc Am 68:81–85

Keirans JE, Clifford CM (1978) The genus *Ixodes* in the United States: a scanning electron microscope study and key to the adults. J Med Entomol Suppl 2:149

Keirans JE, Degenhardt WG (1985) *Aponomma elephense* Price, 1959 (Acari: Ixodidae): diagnosis of the adults and nymph with first description of the larva. Proc Biol Soc Wash 98:711–717

Keirans JE, Clifford CM, Hoogstraal H, Easton ER (1976) Discovery of *Nuttalliella namaqua* Bedford (Acarina: Ixodoidea: Nuttalliellidae) in Tanzania and redescription of the female based on scanning electron microscopy. Ann Entomol Soc Am 69:926–932

Keirans JE, Clifford CM, Reddell JR (1977) Description of the immature stages of *Nothoaspis reddelli* (Ixodoidea: Argasidae) from bat caves in Mexico. Ann Entomol Soc 70:591–595

Keirans JE, Hoogstraal H, Clifford CM (1979) Observations on the subgenus *Argas* (Ixodoidea: Argasidae: *Argas*). 16. *Argas* (*A.*) *moreli*, new species, and keys to the Neotropical species of the subgenus. J Med Entomol 15:157–165

Kohls GM, Hoogstraal H (1960) Observations on the subgenus *Argas* (Ixodoidea, Argasidae, *Argas*). 2. *A. cooleyi*, new species, from western North American birds. Ann Entomol Soc Am 53:625–631

Kohls GM, Hoogstraal H (1961) Observations on the subgenus *Argas* (Ixodoidea, Argasidae, *Argas*). 4. *A. neghmei*, new species, from poultry houses and human habitations in northern Chile. Ann Entomol Soc Am 54:844–851

Kohls GM, Hoogstraal H, Clifford (1961) Observations on the subgenus *Argas* (Ixodoidea, Argasidae, *Argas*). 5. Study of *A. brevipes* Banks, 1908, from birds in Arizona and California, USA, and Baja California, Mexico. Ann Entomol Soc Am 54:869–877

Kohls GM, Sonenshine DE, Clifford CM (1965) The systematics of the subfamily Ornithodorinae (Acarina: Argasidae). II. Identification of the larvae of the Western Hemisphere and descriptions of three new species. Ann Entomol Soc Am 58:331–364

Kohls GM, Hoogstraal H, Clifford CM, Kaiser MN (1970) The subgenus *Persicargas* (Ixodoidea, Argasidae, *Argas*). 9. Redescription and New World records of *Argas* (*P.*) *persicus* (Oken) and resurrection, redescription, and records of *A.* (*P.*) *radiatus* Railliet, *A.* (*P.*) *sanchezi* Duges, and *A.* (*P*). *miniatus* Koch, New World ticks misidentified as *A.* (*P.*) *persicus*. Ann Entomol Soc Am 63:590–606

Kolonin GV (1978) World distribution of ixodid ticks (genus *Haemaphysalis*) Akad Nauk SSSR, Dalln Nauch Tsentr Tikh Inst Geogr Izd "Nauka" Moskva, 71 pp (in Russian)

Kolonin GV (1981) World distribution of ixodid ticks (genus *Ixodes*). Akad Nauk SSSR, Dalln Nauch Tsentr Tikh Inst Geogr Izd "Nauka" Moskva, 114 pp (in Russian)

Kolonin GV (1983) World distribution of ixodid ticks. Genera *Hyalomma*, *Aponomma*, *Amblyomma*. Akad Nauk SSSR, Dalln Nauch Tsentr Tik Inst Geogr Izd "Nauka" Moskva, 120 pp (in Russian)

Kolonin GV (1984) World distribution of ixodid ticks. Genera *Dermacentor*, *Anocentor*, *Cosmiomma*, *Dermacentonomma*, *Nosomma*, *Rhipicentor*, *Rhipicephalus*, *Boophilus*, *Margropus*, *Anomalohimalya*. Akad Nauk SSSR, Dalln Nauch Tsentr Tikh Inst Geogr Izd "Nauka" Moskva, 94 pp (in Russian)

Kusov VN (1973) Ornithodrinae ticks of Kazakhstan and republics of Central Aisa. Akad Nauk Kazakh SSR Inst Zool Alma-Ata 266 pp (in Russian)

Matthysse JG, Colbo MH (1987) The ixodid ticks of Uganda. Entomological Society of America, College Park, MD, 426 pp

Morel PC (1976) Etude sur les Tiques d'Ethiope (Acarines, Ixodides). Ministere des Affaires Etrangeres, Maisons-Alfort, 326 pp (in French)

Neumann LG (1911) Ixodidae. Das Tierreich, No 26 R. Friedlander und Sohn, Berlin, 169 pp

Nunez JL, Munoz-Cobenas ME, Moltedo HL (1985) *Boophilus microplus*. The common cattle tick. Springer, Berlin, Heidelberg New York, 204 pp

Nuttall GHF, Warburton C (1908) On a new genus of the Ixodoidea together with a description of eleven new species of ticks. Proc Camb Philos Soc 14:392–416

Nuttall GHF, Warburton C (1911) Ticks. A monograph of the Ixodoidea. Part II. Ixodidae. Cambridge University Press, Cambridge, pp 105–348

Pegram RG, Clifford CM, Walker JB, Keirans JE (1987a) Clarification of the *Rhipicephalus sanguineus* group (Acari, Ixodoidea, Ixodidae). I. *R. sulcatus* Neumann, 1908 and *R. turanicus* Pomerantsev, 1936. Syst Parasitol 10:3–26

Pegram RG, Keirans JE, Clifford CM, Walker (1987b) Clarification of the *Rhipicephalus sanguineus* group (Acari, Ixodoidea, Ixodidae). II. *R. sanguineus* (Latreille, 1806) and related species. Syst Parasitol 10:27–44

Pegram RG, Walker JB, Clifford CM, Keirans JE (1987c) Comparison of populations of the

Rhipicephalus simus group: *R. simus*, *R. praetextatus*, and *R. muhsamae* (Acari: Ixodidae). J Med Entomol 24:666–682
Pospelova-Shtrom MV (1969) On the system of classification of ticks of the family Argasidae Can., 1890. Acarologia 11:1–22
Roberts FHS (1970) Australian ticks. Commonwealth Scientific and Industrial Organization, Melbourne, 267 pp
Robinson LE (1926) Ticks. A monograph of the Ixodoidea. Part IV. The genus *Amblyomma*. Cambridge University Press, 302 pp
Santos Dias JAT (1985) Previous note concerning the organization of a catalogue for the genus *Aponomma* Neumann, 1899 (Acarina-Ixodoidea). Garcia de Orta Ser Zool 12:31–42
Seneviratana P (1965) The Ixodoidea (ticks) of Ceylon. Parts II and III. Ceylon Vet J 13:28–54
Singh KRP (1968) Description of the nymph and the larva of *Nosomma monstrosum* (Nuttall and Warburton, 1908). Parasitology 58:461–463
Singh KRP, Dhanda V (1965) Description and keys of immature stages of some species of Indian *Hyalomma* Koch, 1844 (Ixodoidea, Ixodidae). Acarologia 7:636–651
Sonenshine DE, Clifford CM, Kohls GM (1962) The identification of larvae of the genus *Argas* (Acarina: Argasidae). Acarologia 4:193–214
Sonenshine DE, Clifford CM, Kohls GM (1966) The systematics of the subfamily Ornithodorinae (Acarina: Argasidae). III. Identification of the larvae of the eastern hemisphere. Ann Entomol Soc Am 59:92–122
Tanskul P(L), Inlao I (1989) Keys to the adult ticks of *Haemaphysalis* Koch, 1844, in Thailand with notes on changes in taxonomy (Acari: Ixodoidea: Ixodidae). J Med Entomol 26:573–600
Tendeiro J (1959) Sur quelques Ixodides du Mozambique et de la Guinee Portugaise. I. Bol Cult Guine Port 14:21–95
Teng K-F (1982) The geographic distribution of the genus *Dermacentor* in China. Sinozoologia (2):211–216
Teng K-F, Huang C-A (1981) A new species of tick of the genus *Anomalohimalaya* from China (Acarina: Ixodidae). Acta Entomol Sin 24:99–102 (in Chinese with English summary)
Theiler G (1961) A contribution to the knowledge of African Ixodidae. The genus *Rhipicentor*. Rev Zool Bot Afr 66:297–308
Theiler G, Robinson BN (1953) Ticks in the South African Zoological Survey collection. Part VII. Six lesser known African rhipicephalids. Onderstepoort J Vet Res 26:93–136
Toumanoff C (1944) Les Tiques (Ixodoidea) de L'Indochine. Inst Pasteur de L'Indochine, Saigon 220 pp (in French)
Trapido H, Varma MGR, Rajagopalan PK, Singh KRP, Rebello MJ (1964) A guide to the identification of all stages of the *Haemaphysalis* ticks of south India. Bull Entomol Res 55:249–270
Walker JB (1961) Some observations on the classification and biology of ticks belonging to the genus *Rhipicephalus*, with special reference to the immature stages. East Afr Med J 38:232–238
Walker JB (1974) The ixodid ticks of Kenya: a review of present knowledge of their hosts and distribution. Commonwealth Institute of Entomology, London, 220 p
Walker JB, Laurence BR (1973) *Margaropus wileyi* sp. nov. (Ixodoidea, Ixodidae), a new species of tick from the reticulated giraffe. Onderstepoort J Vet Res 40:13–21
Walker JB, Keirans JE, Pegram RG, Clifford CM (1988) Clarification of the status of *Rhipicephalus tricuspis* Donitz, 1906 and *Rhipicephalus lunulatus* Neumann, 1907 (Ixodoidea, Ixodidae). Syst Parasitol 12:159–186
Wassef HY, Hoogstraal H (1983) *Dermacentor* (*Indocentor*) *compactus* (Acari: Ixodoidea: Ixodidae): identity of male and female. J Med Entomol 20:648–652
Wassef HY, Hoogstraal H (1984a) *Dermacentor* (*Indocentor*) *auratus* (Acari: Ixodoidea: Ixodidae): identity of male and female. J Med Entomol 21:169–173
Wassef HY, Hoogstraal H (1984b) *Dermacentor* (*Indocentor*) *atrosignatus* (Acari: Ixodoidea: Ixodidae): identity of male and female. J Med Entomol 21:586–591
Wassef HY, Hoogstraal H (1986a) *Dermacentor* (*Indocentor*) *taiwanensis* (Acari: Ixodoidea: Ixodidae): identity of male and female. J Med Entomol 23:173–177
Wassef HY, Hoogstraal H (1986b) *Dermacentor* (*Indocentor*) *steini* (Acari: Ixodoidea: Ixodidae): identity of male and female. J Med Entomol 23:532–537

Wassef HY, Hoogstraal H (1988) *Dermacentor* (*Indocentor*) *steini* (Acari: Ixodoidea: Ixodidae): hosts, distribution in the Malay Peninsula, Indonesia, Borneo, Thailand, the Philippines, and New Guinea. J Med Entomol 25:315–320

Yamaguti N, Tipton VJ, Keegan HL, Toshioka S (1971) Ticks of Japan, Korea, and the Ryukyu Islands. Brigham Young Univ Sci Bull Biol Ser 15:1–226

Yeoman GH, Walker JB (1967) The ixodid ticks of Tanzania: a study of the zoogeography of the Ixodidae of an east African country. Commonwealth Institute of Entomology, London, 215 p

Yunker CE, Keirans JE, Clifford CM, Easton ER (1986) *Dermacentor* ticks (Acari: Ixodiodea: Ixodidae) of the New World: a scanning electron microscope atlas. Proc Entomol Soc Wash 88:609–627

Zumpt F (1949) Preliminary study to a revision of the genus *Rhipicephalus* Koch. Mocambique 15:57–169

Veterinary Significance of Ticks and Tick-Borne Diseases

G. UILENBERG[1]

Summary

The veterinary importance of ticks and tick-borne diseases, with the emphasis on cattle, small ruminants, horses and dogs in the tropics and subtropics, is reviewed. It is stressed that the importance in a given region depends not only on the local tick species and the diseases they transmit, but to a large extent on the composition of the livestock population. Local breeds are more resistant to tick infestation and to many of the tick-borne diseases than exotic and cross-bred animals.

Tick infestation by itself can cause major problems such as anaemia, loss in production, abscesses leading to loss of teats and lameness, as well as myiasis. In humid conditions African *Amblyomma* ticks are associated with severe forms of dermatophilosis in susceptible cattle. Losses to the hide industry may be considerable, while tick toxicoses are important in some areas.

The main diseases transmitted by ticks to livestock are anaplasmosis (ruminants), babesiosis (ruminants, horses, dogs), theileriosis (ruminants, horses) and cowdriosis (ruminants). Ehrlichiosis in ruminants and dogs is also important in certain tropical and subtropical regions. Endemic stability can often be achieved, especially in indigenous livestock and even East Coast fever may cause no more than a slightly increased calf mortality in local zebu in fully endemic areas.

Estimates of economic losses due to ticks and tick-borne diseases are often little more than educated guesses. Any form of control in local resistant livestock is not always cost-effective, whereas intensive and expensive control measures are often required for valuable exotic breeds. On a global scale, ticks are undoubtedly the most important ectoparasites of livestock.

Smith and Kilborne (1893) appear to have been the first to demonstrate transmission of a disease, bovine babesiosis, by an arthropod vector, *Boophilus annulatus*, during their investigation of Texas fever. (This fact must have given many a veterinary parasitologist a secret feeling of superiority over his or her medical colleagues . . .) Although cattle owners in the southern USA had connected Texas fever with the appearance of ticks, the official view had been that "The tick theory has acquired quite a renown during the past summer, but a little thought should have satisfied any one of the absurdity of the idea . . . " (John Gamgee, in: Report of the Commissioner of Agriculture on the diseases of cattle in the United States, 1871, quoted by Smith and Kilborne 1893).

Since that time the importance of ticks and tick-borne diseases of domestic animals has been increasingly realized and no one who knows anything on the subject doubts their veterinary significance. There is probably no other group of

[1]Institut d'Elevage et de Médecine Vétérinaire des Pays Tropicaux, 10, Rue Pierre Curie, 94704 Maisons-Alfort, France

arthropods which transmit such a variety of infective organisms to animals and man, ranging from viruses through bacteria and protozoa even to helminths, many of which are the cause of important diseases. But ticks themselves can also be harmful without acting as vectors of disease.

It is not my intention to try in this chapter to review systematically and in detail the literature and all aspects of damage and diseases caused by ticks, but rather to give a general impression of the subject and highlight some examples. The emphasis will be on livestock in the warmer regions of the world, the tropics and subtropics, where ticks and tick-borne diseases are most numerous and most important. In some areas the importance of the diseases they transmit exceeds that of many of the major contagious and other vector-borne diseases. Much, however, depends on the circumstances, not only on the local species of ticks and tick-borne disease agents, but to a very large extent also on the degree of susceptibility of the livestock kept in the area, to infestation by these ticks and the diseases transmitted. For instance in Kenya, East Coast fever (caused by *Theileria parva*) is often considered to be the most important disease of cattle. But this is only so because of the presence of a large susceptible population of exotic and cross-bred animals. East Coast fever is certainly not a major disease for local East African zebu cattle in endemic areas, as long as there is no tick control to upset endemic stability; mortality in young calves of this breed in endemic areas is low and all the older animals have acquired immunity at an early stage. The breed has an innate resistance, presumably the result of natural selection over many generations. Genetically determined resistance to tick-borne diseases in local breeds of livestock is a common and general feature of the epidemiology of these diseases. The same is true for tick infestation, particularly where one-host *Boophilus* spp. are concerned. Uncontrolled *B. microplus* infestations will severely affect most European *Bos taurus* cattle and may even kill them, having little effect, if any, on pure zebu cattle. *Bos indicus* cattle have a far greater genetically determined ability to limit the infestation by developing an effective defensive response, which is mainly immunological.

A unique example of the problems that ticks per se can present is provided by the situation on New Caledonia, the French territory in the Pacific Ocean. It appears that an American officer stationed on the island during the second World War conceived the idea of importing mules, because of the mountainous nature of the land. The mules were purchased in Australia, and introduced the tick *Boophilus microplus* on New Caledonia, towards 1942. Fortunately they did this without bringing along the important cattle diseases of which this tick is a vector, babesiosis (both *Babesia bigemina* and *B. bovis*) and anaplasmosis (Rageau and Vervent 1959). However, the tick found an ideal environment in which to prosper: a very favourable climate for its development throughout the year on most of the island and hosts with little innate capability for mounting an effective immune response to the infestation, since most of the cattle on New Caledonia belong to European beef breeds (Limousin, Charolais and various crosses). The tick has created a major problem for the livestock industry on the island, where these cattle can only be productive if submitted regularly to acaricidal treatment. The alternative is a tremendous loss in production, and even deaths from tick infestation are known to occur in untreated animals. The problem is, at present, compounded by widespread acaricide resistance, the unavoidable result of intensive acaricidal tick control. Resistance to an organophosphorous compound (ethion), in use since the early 1970s, has developed in many parts of the island, while resistance to DDT (used from 1947 to 1973) has also been shown to exist (Brun et al. 1984). This may complicate the use of pyrethroids in place of organophosphorous acaricides (Nolan et al. 1977). The prospects for successful beef production using European cattle in New Caledonia have declined since the discovery of strains of *B. microplus* in Australia with high levels of resistance to pyrethroids and amidines (Nolan 1990). It is hoped that the tick vaccine being tested in Australia (Kemp et al. 1989) will become an effective and economic reality.

(We may add that a fatal case of bovine babesiosis has just been observed on New Caledonia, in a herd of dairy cattle partly consisting of animals imported from Australia (P.C. Morel and G. Uilenberg unpubl. observ. Jan. 1990). A definitive species identification was difficult because of post-mortem changes, but the smears appeared to contain both *B. bigemina* and *B. bovis*. At the time of writing it is not yet known whether the infection can still be eradicated or has spread beyond control.)

The direct effect of *Boophilus* infestation is proportional to the numbers of successfully engorging ticks on the animal. Heavy infestations on susceptible cattle cause weight loss, anaemia and mortality (Corrier et al. 1979), mainly because of loss of blood, although other effects have also been shown to occur (summarized by Springell 1974 and Taylor and Plumb 1981). The populations of these ticks can increase to enormous densities in the presence of susceptible (taurine) cattle and a suitable climate. On the other hand, even in a favourable climate for the tick, populations will remain limited if the cattle are tick-resistant. The average cattle owner on New Caledonia would therefore probably be well advised to turn towards better adapted cattle breeds. A number of Santa Gertrudis are already kept on the island and could certainly be selected for higher resistance. Bourne et al. (1988) conclude from their study that the resistance of zebu-type cattle to *B. microplus*, combined with their advantages in productivity in (the Australian) tropical and subtropical regions, means that there is no apparent reason to retain *Bos taurus* beef breeds in the region and spend large sums on tick control.

Soft ticks, Argasidae, may also be economically important, because of the direct loss of blood they can cause. The effects of ticks of the *Argas persicus* group on poultry, and of *Ornithodoros savignyi* on camels and cattle are well-known. A vivid description of the attack by the latter on cattle has been given by Hoogstraal (1956).

However, there are other ways in which ticks per se affect the production of livestock, including that of animals belonging to tropical breeds. The bite of large ticks with a long, massive hypostome, such as the *Amblyomma* spp., is commonly followed by abscesses due to secondary bacterial infection. Depending on the site, such abscesses may have serious consequences such as loss of teats and quarters, or lameness if the tick has attached in the interdigital space.

Amblyomma hebraeum appears to be particularly important in this respect, at least in Mozambique, where Asselbergs and Lopes Pereira (1990) report on field studies carried out in 1983 and 1984 among small farmers with zebu cattle, where dipping schemes had broken down. In 1983 only 67% of adult cows had 4 functional teats and 19% had only 2 teats or less left; in 1984 these figures were 53% and 26% respectively and in a herd of 117 heifers only 48% had 4 undamaged teats. In 1984, mortality in calves of cows with 4 teats was 18%, with 3 teats 19%, with 2 teats 44%, with 1 teat 86%, and in calves of cows without functional teats, of course, 100%. The main cause of the damage was *A. hebraeum*, often found in large clusters on the udder. In 1988, in a herd with smaller numbers of the tick because of better dip management, still only 66% of the cows had undamaged udders and 10% had two teats or fewer left.

Another way in which the large African *Amblyomma* species are harmful is through their apparent immunodepressive effect. Severe bovine dermatophilosis (a skin disease caused by *Dermatophilus congolensis*, a virtually cosmopolitan bacterium) occurs mainly in exotic cattle (meaning not indigenous to the endemic area), particularly in association with *Amblyomma variegatum*, although other factors such as humidity are also important. This is spectacularly evident in the Caribbean area where *A. variegatum* has been introduced from Africa. On those islands where the tick has managed to establish itself, dermatophilosis immediately becomes the major disease problem in cattle (Butler 1975; Burridge et al. 1984; Uilenberg et al. 1984). If the tick is not intensively controlled cattle breeding may become uneconomic. The way in which the presence of the tick was discovered on the island of Dominica is worth relating. Most of the eastern Caribbean islands were inspected for the presence of *A. variegatum* in the period between 1982 and 1984. In the course of this survey, the island of Dominica, where the tick was not known to occur, was visited in December 1983 by a team consisting of M.J. Burridge, E.F. Birnie and the author. During a discussion with the veterinary staff, mention was made of a new and intractable skin disease of cattle on a farm in the interior. Birnie was taken to the farm and found dermatophilosis as well as *A. variegatum*, which apparently occurred only locally there. The disease thus acted as a marker for the presence of the tick! The tick has not been shown to be a biological vector of *D. congolensis*, but in one way or another it breaks down the host's defenses against this micro-organism; it is most likely that its saliva is immunodepressive. Immunodepression by tick infestation has been described (summary by Wikel and Whelen 1986) and the immunodepressive effects of tick saliva have been reported (Stewart 1983; Ribeiro et al. 1985).

Infestation by ticks, even those with short mouthparts, can also induce other complications, such as myiasis by *Chrysomya bezziana*. This is especially common in Zimbabwe in the ears of cattle infested by *Rhipicephalus appendiculatus* (Norval et al. 1988). One shudders to think what might happen if the American screw-worm fly, *Cochliomyia hominivorax*, which has recently been introduced into North Africa, managed to reach sub-Saharan Africa! In the USA, it is a serious complication of infestation of cattle by *Amblyomma maculatum* (Ahrens et al. 1977).

Certain ticks are able to cause disease by toxins in their saliva. The toxins of some ticks induce paralysis, which can be fatal even in adult cattle in the case of *Dermacentor andersoni* in Canada and the USA. The saliva of *Hyalomma truncatum* (but only of certain strains) causes "sweating sickness", a generalized disease associated with an eczema-like skin condition. A general review of the tick toxicoses is given by Gothe (1981).

Ticks also cause considerable losses to livestock owners because of the decrease in value of the damaged hides and skins. Even ticks with short mouth parts, such as *Boophilus*, depreciate the value of the leather. In Australia, where *B. microplus* is the main cattle tick, "ticks are the biggest problem facing the hide industry in Queensland" (Daniel and McGuiness 1977) and the depreciation in the value of hides contributes 5% to the cost of the tick problem in Australia (Springell 1974).

The veterinary significance of tick-borne diseases in most areas of the world is much greater than that of the ticks themselves. The main diseases are mentioned below, but only those which are important to animals, not those for which ticks are merely reservoirs of infection for man (such as Lyme disease, certain rickettsial infections, many of the arboviruses, etc.). Ticks and tick-borne diseases are probably of the greatest economic importance for cattle. Only cattle (including domestic buffalo, which probably have all the infections that affect cattle), sheep, goats, horses and dogs are covered here, but all domestic animals are affected, including pigs and poultry.

Anaplasmosis of cattle (mainly *Anaplasma marginale*) is transmitted by several tick genera and species, and to some extent also mechanically by biting insects, in almost all tropical and subtropical regions, where it is regarded as one of the most important diseases of cattle. *Boophilus* spp. are among the main vectors. Endemic stability can be achieved because of an inverse age resistance of calves.

Cattle are affected by babesiosis in temperate as well as tropical and subtropical regions. *Babesia bigemina* and *B. bovis* occur in the tropics and subtropics, with *Boophilus* ticks as vectors; *B. bovis* is particularly pathogenic for European cattle. Endemic stability is possible in areas with sufficient infected ticks, again because of resistance in calves, which may possibly be increased by passive immunity. In Europe and temperate Western Asia *B. divergens* (transmitted by *Ixodes* spp.) and to a lesser extent *B. major* (with *Haemaphysalis punctata* as its vector) are of some importance. The latter is replaced in the Far East by *B. ovata*, also transmitted by *Haemaphysalis* ticks.

Cowdriosis, or "heartwater", caused by *Cowdria ruminantium*, a rickettsial organism like *Anaplasma* and *Ehrlichia*, occurs in almost all of sub-Saharan Africa, where it is transmitted by *Amblyomma* ticks. It is a major cause of mortality in exotic breeds, but indigenous cattle in endemic areas are usually quite resistant. This is one of the main reasons why the disease remained undiscovered in many African countries until the importation of exotic cattle. The disease also occurs on a few Caribbean islands, where it was introduced in the 19th century with one of its African tick vectors, and constitutes a threat to the American mainland.

Dermatophilosis, although not strictly speaking a tick-borne disease, has already been discussed as a major cause of disease in exotic cattle in association with the presence of large African *Amblyomma* ticks.

Ehrlichiosis is of some importance in temperate Europe, where *Ixodes ricinus* transmits *Ehrlichia phagocytophila*. *E. ondiri* is locally important in East Africa, but its vector is still unknown, while *E. bovis* has been reported as a cause of disease and mortality in West Africa, where it is possibly transmitted by *Hyalomma* spp. (Mild strains of *E. bovis* are transmitted in eastern and southern Africa by *Rhipicephalus appendiculatus*.)

Theileriosis is particularly important in eastern, central and southern Africa, where *Theileria parva* is transmitted by ticks of the *Rhipicephalus appendiculatus* group, and in the Mediterranean Basin, the Near and Middle East, and central Asia, where *Hyalomma* ticks are vectors of *T. annulata*. *Theileria parva* and *T. annulata* are among the most important cattle diseases in the areas concerned, especially in "exotic" and cross-bred cattle. Indigenous populations, having acquired a high degree of resistance through selection pressure over many generations, may live with the disease in a state of endemic stability, but even these cattle are likely to die if they acquire infection as adults. The pathogenic significance of *T. mutans* (with *Amblyomma* ticks as vectors), in sub-Saharan Africa and the Caribbean is probably negligible in most cases, but a primary infection in adult cattle may cause severe anaemia. Some strains may be more pathogenic than others. *T. taurotragi*, transmitted by *Rhipicephalus* ticks in the same regions as *T. parva*, is normally benign, but may sometimes be a cause of "turning sickness" or cerebral theileriosis. Finally, *T. orientalis* (for which the correct name may possibly be *T. buffeli*), an almost cosmopolitan species transmitted by *Haemaphysalis* ticks, is usually mild, but more pathogenic strains are known from eastern Asia. In Africa where the vector is as yet unknown, it has been diagnosed in association with disease outbreaks although it has not been proven that it caused the disease (Becerra et al. 1983; Kiltz et al. 1986).

Small ruminants are also affected by anaplasmosis (due to *Anaplasma ovis*) in most tropical and subtropical areas. Knowledge of its vectors, epidemiology, and pathological significance is lacking. Babesiosis is also common in tropical and subtropical areas of the Old World. *Babesia motasi* and *B. ovis*, transmitted by *Haemaphysalis* spp. and *Rhipicephalus bursa* respectively, cause disease in the Mediterranean basin and the Near East. Similar, but poorly studied, parasites occur elsewhere as well, including sub-Saharan Africa, where the vectors remain unknown. *Cowdria ruminantium* infection, already discussed under cattle diseases, is also of great importance for small ruminants. Here again, cowdriosis is of particular significance for "exotic" stock with extremely high mortalities for instance in Angora goats. Indigenous breeds may also suffer considerably in certain circumstances. Ehrlichiosis in Europe due to *Ehrlichia phagocytophila* (see under cattle) may cause immunodepression and abortions. As yet tropical ehrlichiosis in small ruminants has not been considered as an important disease. Recently though, *E. ovina* was associated with considerable mortality in adult Sahelian sheep in Senegal exposed to natural tick infestation in a more humid coastal area (Gueye et al. 1989). Theileriosis as a disease is mainly limited to the Mediterranean basin and the Near and Middle East, where *Theileria lestoquardi* (formerly *T. hirci*) is transmitted by *Hyalomma* ticks and causes high mortality in sheep and goats. It is noteworthy that strains of tick-borne blood parasities of sheep and goats often shown considerable differences in their infectivity and pathogenicity for these two species of small ruminants.

Horses are affected by babesiosis (*Babesia caballi*) and theileriosis (*Theileria equi*, formerly *Babesia equi*) in most of the tropics and subtropics. Vectors are ticks belonging to the genera *Dermacentor*, *Hyalomma* and *Rhipicephalus*. Apart from their direct importance as disease-causing agents, they are an important obstacle to free international movements of equines, especially *T. equi*, which is virtually impossible to eradicate in carrier animals.

Finally, tick-borne diseases are of considerable importance in dogs. *Rhipicephalus sanguineus* is the vector, throughout the tropics and subtropics, of babesiosis (*Babesia canis vogeli* and *B. gibsoni*), of ehrlichiosis (*Ehrlichia canis*) and of hepatozoonosis (*Hepatozoon canis*). In temperate regions of Europe and probably Asia *Dermacentor* spp. transmit babesiosis due to *B. canis canis*, while *Haemaphysalis* ticks are the vectors of the very virulent *B. canis rossi* in Africa (Uilenberg et al. 1989).

It is not always realized, even by veterinarians, that the significance of tick-borne diseases of domestic animals varies considerably according to the circumstances. The possibility of achieving endemic stability for some of the tick-borne diseases, and the importance of genetically determined differences between indigenous and exotic livestock, cannot be sufficiently stressed.

The attitude of the governmental veterinary services and the history of the control of ticks and tick-borne diseases in much of Africa has been influenced for a long time by the traumatic events in southern Africa, where East Coast fever caused by *Theileria parva parva* was introduced with cattle from East Africa (Henning 1956; Lawrence 1982). This followed the pan-African rinderpest epidemic at the end of the 19th century, which had decimated the cattle populations. The enormous losses resulted in a great demand for the importation of cattle, both for providing beef and for restocking purposes. In 1901 a shipload of cattle from Dar-es-Salaam introduced East Coast fever into Mozambique, Zimbabwe and South Africa. The disease proved to be extremely virulent for all types of cattle in southern Africa, local breeds as well as those of European origin. Within a few years half of the few cattle still owned by ranchers in Zimbabwe had died from this new scourge, while in South Africa the losses were counted by the hundreds of thousands.

Drastic government measures were taken in southern Africa, including restrictions of cattle movements, compulsory monitoring of all deaths by the submission of smears, quarantining, compulsory short interval dipping and finally wholesale slaughter of infected herds and starving out infected ticks on pastures. Although the vector was not eradicated, classical East Coast fever had disappeared in Zimbabwe and South Africa by 1955.

Eradication of the disease in eastern and central Africa has not been possible because the situation there is very different. Several factors have played an important role, such as the reservoir role of recovered cattle (carrier state), the impossibility of enforcing the necessary measures because of financial, organizational, political and other constraints, including acaricide resistance, etc. The only applicable method of control, without which it was impossible to rear exotic cattle in East Coast fever areas, was short-term dipping. Intensive acaricidal treatment, often compulsory, was blindly applied even to indigenous zebu populations. However, in East Africa these have presumably lived with the disease for generations and careful observations have later shown that, in the absence of tick control, East Coast fever may cause no more than a slightly increased calf mortality in fully endemic areas (Barnett 1957; Dolan and Young 1981; Moll et al. 1984). Nevertheless, theilerial infections may well have a considerable negative influence on the weight gain in calves of local breeds in endemic areas (Moll et al. 1986). Of course, the situation is quite different at the fringe of the endemic zone, where there is no endemic stability (McCulloch et al. 1968; Yeoman 1969).

Evaluations of the economic impact of ticks and tick-borne diseases are often little more than educated guesses, based on unreliable estimates and assumptions. Factors that have to be taken into account include:- loss of production through diminished growth rate or weight loss, a decline in milk yield and diminished value of hides. Mortality and all aspects of the cost of tick control and of the cost of control of tick-borne diseases per se need to be considered. When control is improved it usually results in higher costs, although the loss of production and the mortality will decline. Therefore, whether the higher cost of control is economic depends on the circumstances. In valuable, imported, susceptible breeds losses caused by ticks and tick-borne diseases are usually so high that intensive, expensive control measures are cost-effective, whereas any form of control may not always be economic where local resistant populations are concerned.

In the introduction to a field manual, FAO (1984) has estimated total annual world losses due to ticks and tick-borne diseases at roughly US$7000 million, but adds that the total losses due to ticks may in fact be much higher. The summarised results of a few studies on the economic impact of ticks, either by themselves or in combination with the diseases they transmit, are given only as examples, and in no case should any figures quoted be assumed to be completely reliable. Wide variations exist between various estimates of the effects of ticks, even on the same pastures, which can partly be explained by seasonal influences (Sutherst et al. 1983). General reviews are, for instance, given by Steelman (1976) and Gibbs (1985). An official commission concluded that in 1973 the presence in Australia of the cattle tick *B. microplus* and the diseases it transmits was costing the cattle producers and the veterinary services US$62 million per year (Springell 1974). De Castro et al. (1985) found significant differences in tick infestation and liveweight gain between dipped and undipped Boran (zebu) heifers. They concluded, though, that economic thresholds for tick control remained to be established, and that such cattle can be kept in theileriosis endemic areas (if immunized against East Coast fever) without acaricide application, at the expense of some loss in productivity. Norval et al. (1988), using controlled single species infestations, have calculated a reduction in liveweight gain in 18 to 24 months old cattle of roughly 4 g per engorging female *Rhipicephalus appendiculatus* (not counting the influence of screw-worm infestations which were immediately treated). Liveweight gains of heavily infested European breed cattle were severely affected. On resistant Sanga animals, though, the effect was insignificant.

In similar experiments it was found that each engorging female of *Amblyomma hebraeum* caused a loss in liveweight gain of some 10 g (Norval et al. 1989).

Larvae and nymphs of both these tick species had no significant effect on liveweight gain.

Studies on the influence of natural tick infestations in Zambia indicate a much larger loss in live weight gain caused by *Amblyomma variegatum* (45–60 g per engorged female) (Pegram et al. 1989). Nevertheless, Pegram and Chizyuka (1990) found that intensive tick control in herds of indigenous cattle exposed to natural infestation costs more than the benefit derived from it, but seasonal strategic control may be economically justified.

It has been estimated that the presence of *Amblyomma variegatum* on Guadeloupe means an annual economic loss for the 75 000 cattle on the island of some US$850 000 (Camus and Barré 1990). Of this 50% is due to dermatophilosis, 17% to heartwater and 33% to the ticks themselves. To this loss should be added the cost of acaricidal treatment, estimated at $700 000. Potential economic losses, in the absence of tick control, have been estimated at $4.6 million per year. These figures certainly do not do justice to the importance of this tick in the Caribbean. On Guadeloupe the very resistant Creole cattle constitute 91.4% of the catttle population and potentially more productive imported breeds only 1%, the rest being crosses (Salas et al. 1988). This composition of the cattle population is peculiar to Guadeloupe, where the tick, cowdriosis and dermatophilosis have maintained a selective pressure for 150 years. Under these circumstances the Creole zebu has shown itself to be the best and most productive choice. The havoc caused by the recent introduction of *A. variegatum* on many of the other islands in the Caribbean is far greater, because of the presence of cattle populations which are highly susceptible to dermatophilosis. Its greatest importance, though, lies in the constant threat it presents to the American mainland (Barré et al. 1987).

It is hoped that this chapter conveys some idea of the veterinary significance of ticks. On a global scale they are undoubtedly the most important ectoparasites of livestock.

References

Ahrens EH, Gladney WJ, McWhorter GM, Deer JA (1977) Prevention of screwworm infestation in cattle by controlling Gulf Coast ticks with slow release insecticide devices. J Econ Entomol 70:581–585

Asselbergs MJ, Lopes Pereira CM (1990) Damage by *Amblyomma hebraeum* in local *Bos indicus* cattle in Mozambique. In: Kuil H, Paling RW, Huhn JE (eds) Livestock production and diseases in the tropics. Proc 6th Int Conf Inst Trop Vet Med, Wageningen, 28 Aug–1 Sept 1989, pp 358–360

Barnett SF (1957) Theileriasis control. Bull Epizoot Dis Afr 5:343–357

Barré N, Uilenberg G, Morel PC, Camus E (1987) Danger of introducing heartwater onto the American mainland: potential role of indigenous and exotic *Amblyomma* ticks. Onderstepoort J Vet Res 54:405–417

Becerra VM, Eggen AAS, De Rooy RC, Uilenberg G (1983) *Theileria orientalis* in cattle in Ethiopia. Res Vet Sci 34:362–364
Bourne AS, Sutherst RW, Sutherland ID, Maywald GF, Stegeman DA (1988) Ecology of the cattle tick (*Boophilus microplus*) in subtropical Australia. III. Modelling populations on different breeds of cattle. Aust J Agric Res 39:309–318
Brun LO, Wilson JT, Nolan J (1984) Patterns of resistance in five samples of ethion-resistant cattle tick (*Boophilus microplus*) from New Caledonia. Trop Pest Manage 30:296–301
Burridge MJ, Barré N, Birnie EF, Camus E, Uilenberg G (1984) Epidemiological studies on heartwater in the Caribbean, with observations on tick-associated bovine dermatophilosis. Proc 13th World Congr Diseases of Cattle, Durban, pp 542–546
Butler MC (1975) Dermatophilosis of cattle, sheep and goats in St. Lucia. State Vet J 30:279–283
Camus E, Barré N (1990) *Amblyomma variegatum* and associated diseases in the Caribbean. Strategies for control and for eradication in Guadeloupe. Parassitologia (Rome) 32:185–193
Corrier DE, Vizcaino O, Terry M, Betancourt A, Kuttler KL, Carson CA, Trevino G, Ristic M (1979) Mortality, weight loss and anaemia in *Bos taurus* calves exposed to *Boophilus microplus* ticks in the tropics of Colombia. Trop Anim Health Prod 11:215–221
Daniel DJ, McGuiness PJ (1977) Facts on cattle hides. Queensland Agric J 103:109–112
De Castro JJ, Young AS, Dransfield RD, Cunningham MP, Dolan TT (1985) Effects of tick infestation on Boran (*Bos indicus*) cattle immunised against theileriosis in an endemic area of Kenya. Res Vet Sci 39:279–288
Dolan TT, Young AS (1981) An approach to the economic assessment of East Coast fever in Kenya. In: Irvin AD, Cunningham MP, Young AS (eds) Advances in the control of theileriosis. Martinus Nijhoff, The Hague, pp 412–415
FAO (1984) Ticks and tick-borne disease control. A practical field manual. I. Tick control. FAO Rome, 320 pp
Gibbs HC (1985) Effects of parasites on animal and meat production. In: Gaafar SM, Howard WE, Marsh RE (eds) Parasites, pests and predators. Elsevier, Amsterdam, pp 7–27
Gothe R (1981) Tick toxicoses of cattle. In: Ristic M, McIntyre I (eds) Diseases of cattle in the tropics. Economic and zoonotic relevance. Martinus Nijhoff, The Hague, pp 587–598
Gueye A, Mbengue M, Diouf A, Vassiliades G (1989) Prophylaxie de la cowdriose et observations sur la pathologie ovine dans la région des Niayes au Sénégal. Rev Elev Méd Vét Pays Trop 42:497–503
Henning MW (1956) Animal diseases in South Africa. 3rd edn, Central News Agency, South Africa, pp 593–596
Hoogstraal H (1956) African Ixodoidea, vol I. Ticks of the Sudan. Res Rep NM 005 050.29.07, US Naval Medical Research Unit No 3, Cairo, p 197
Kemp DH, Pearson RD, Gough JM, Willadsen P (1989) Vaccination against *Boophilus microplus*: localization of antigens on tick gut cells and their interaction with the host immune system. Exp Appl Acarol 7:43–58
Kiltz HH, Uilenberg G, Franssen FFJ, Perié NM (1986) *Theileria orientalis* occurs in central Africa. Res Vet Sci 40:197–200
Lawrence JA (1982) East Coast fever: one of the great cattle plagues of Africa. Zimbabwe Sci News 16:14–18
McCulloch B, Suda B' QJ, Tungaraza R, Kalaye WJ (1968) A study of East Coast fever, drought and social obligations, in relation to the need for the economic development of the livestock industry in Sukumaland, Tanzania. Bull Epizoot Dis Afr 16:303–326
Moll G, Lohding A, Young AS (1984) Epidemiology of theilerioses in the Trans-Mara Division, Kenya: husbandry and disease background and preliminary investigations on theilerioses in calves. Prev Vet Med 2:801–831
Moll G, Lohding A, Young AS, Leitch BL (1986) Epidemiology of theileriosis in calves in an endemic area of Kenya. Vet Parasitol 19:255–273
Nolan J (1990) Acaricide resistance in single and multi-host ticks and strategies for its control. Parassitologia (Rome) 32:145–153
Nolan J, Roulston WJ, Wharton RH (1977) Resistance to synthetic pyrethroids in a DDT-resistant strain of *Boophilus microplus*. Pestic Sci 8:484–486
Norval RAI, Sutherst RW, Kurki J, Gibson JD, Kerr JD (1988) The effect of the brown ear-tick

Rhipicephalus appendiculatus on the growth of Sanga and European breed cattle. Vet Parasitol 30:149–164

Norval RAI, Sutherst RW, Jorgensen OG, Gibson JD, Kerr JD (1989) The effect of the bont tick (*Amblyomma hebraeum*) on the weight gain of Africander steers. Vet Parasitol 33:329–341

Pegram RG, Chizyuka HGB (1990) The impact of natural infestations of ticks in Zambia on the productivity of cattle and implications for tick control strategies in central Africa. Parassitologia (Rome) 32:165–175

Pegram RG, Lemche J, Chizyuka HGB, Sutherst RW, Floyd RB, Kerr JD, McCosker PJ (1989) Effect of tick control on liveweight gain of cattle in central Zambia. Med Vet Entomol 3:313–320

Rageau J, Vervent G (1959) Les tiques (Acariens Ixodoidea) des Iles françaises du Pacifique. Bull Soc Pathol Exot 52:819–834

Ribeiro JMC, Makoul GT, Levine J, Robinson DR, Spielman A (1985) Antihemostatic, antiinflammatory, and immunosuppressive properties of the saliva of a tick, *Ixodes dammini*. J Exp Med 161:332–344

Salas M, Planchenault D, Roy F (1988) Etude des systèmes d'élevage bovin traditionnel en Guadeloupe, Antilles françaises. Résultats d'enquêtes. Rev Elev Méd Vét Pays Trop 41:197–207

Smith T, Kilborne FL (1893) Investigations into the nature, causation, and prevention of Texas or southern cattle fever. Bull. No 1, USDA, Bureau of Animal Industry, Washington, pp 301

Springell PH (1974) The cattle tick in relation to animal production in Australia. World Anim Rev (10):19–23

Steelman CD (1976) Effects of external and internal arthropod parasites on domestic livestock production. Annu Rev Entomol 21:155–178

Stewart CG (1983) In vitro immunosuppression of bovine mononuclear cells by tick saliva. S Afr J Sci 79:119

Sutherst RW, Maywald GF, Kerr JD, Stegeman DA (1983) The effect of cattle tick (*Boophilus microplus*) on the growth of *Bos indicus* × *B. taurus* steers. Aust J Agric Res 34:317–327

Taylor RJ, Plumb IR (1981) The effect of natural tick infestation on various blood components and livemass in the bovine in South Africa. In: Whitehead GB, Gibson JD (eds) Tick biology and control. Tick Research Unit, Rhodes University, Grahamstown, pp 21–28

Uilenberg G, Barré N, Camus E, Burridge MJ, Garris GI (1984) Heartwater in the Caribbean. Prev Vet Med 2:225–267

Uilenberg G, Franssen FFJ, Perié NM, Spanjer AAM (1989) Three groups of *Babesia canis* distinguished and a proposal for nomenclature. Vet Q 11:33–40

Wikel SK, Whelen AC (1986) Ixodid-host immune interaction. Identification and characterization of relevant antigens and tick-induced host immunosuppression. Vet Parasitol 20:149–174

Yeoman GH (1969) A field study of a tick vector/cattle disease relationship in East Africa. In: Evans GO (ed) Proc 2nd Int Congr Acarol 1967. Akadémiai Kiado, Budapest, pp 589–594

Economics, Epidemiology and Ecology: A Multidisciplinary Approach to the Planning and Appraisal of Tick and Tick-Borne Disease Control in Southern Africa

R.A.I. Norval[1], J.C. Barrett[2], B.D. Perry[3] and A.W. Mukhebi[3]

Summary

For 75 years acaricide treatment of cattle has been widely practised in southern Africa for the control of ticks and tick-borne diseases (TBDs). This paper discusses the need to reappraise veterinary strategy in this area, because of:

- widespread development of acaricide resistance in ticks;
- evidence that present strategies are not cost-effective;
- the financial burden of intensive dipping on farmers who themselves bear the cost of such dipping; and
- financial, infrastructural and institutional constraints on veterinary programmes funded by Governments in some countries in southern Africa.

The prospect of alternative approaches to tick and TBD control is discussed in relation to:

- the existence of endemic stability for the major TBDs in southern Africa;
- the availability of vaccines for control of ticks and TBDs;
- advances in our understanding and modelling of TBD epidemiology, tick ecology and veterinary economics.

New strategies must reflect fully current knowledge of tick ecology, the epidemiology of TBDs and the economic consequences of ticks and TBDs for livestock productivity. The scope in Zimbabwe for moving away from intensive dipping of cattle is briefly discussed, as an example of the multidisciplinary approach, which is considered appropriate for the design and appraisal of new veterinary strategies. Finally, the paper considers broader economic, veterinary, social, political and institutional aspects of major changes in policy towards tick and TBD control.

1 Introduction

Acaricide treatment of cattle has been practised widely in southern Africa[4] for 75 years, since cattle were first dipped in arsenic oxide for control of East Coast fever (ECF), caused by *Theileria parva*, which had been introduced from eastern Africa at the turn of the century (Lawrence 1992). Tick control played an

---

[1] Center for Tropical Animal Health, College of Veterinary Medicine, University of Florida, Gainesville, Florida 32610, USA
[2] Overseas Development Natural Resources Institute, Central Avenue, Chatham Maritime, Chatham, Kent ME4 4TB, England
[3] International Laboratory for Research on Animal Diseases, P.O. Box 30709, Nairobi, Kenya
[4] Southern Africa refers to Africa south of the Zambezi river, including southern Mozambique, Zimbabwe, Botswana, Namibia, Swaziland, Lesotho and the Republic of South Africa, including its autonomous 'homelands'

important role in the control of ECF and contributed to its final disappearance from South Africa, Swaziland, Zimbabwe and Mozambique by 1960 (Norval et al. 1990a; Lawrence 1992). Acaricide treatment has contributed greatly to the control of other tick-borne diseases (TBDs), including:

- heartwater, caused by *Cowdria ruminantium*;
- babesiosis, caused by *Babesia bovis* and *B. bigemina*;
- anaplasmosis, caused by *Anaplasma marginale*;
- January disease and corridor disease, caused by *Theileria parva.*

Furthermore, tick control provided a means of preventing direct tick damage and associated production losses, as well as secondary infestation with screw worm fly (*Chrysomya bezziana*) larvae, hide damage, tick associated paralysis and dermatophilosis caused by *Dermatophilus congolensis.*

On the one hand, tick control using acaricides has proved a simple and effective veterinary procedure. On the other hand, it may cause epidemiological problems and is becoming increasingly costly. The development of alternative approaches, such as the use of tick resistant cattle and the possibility of effective immunisation against TBDs, has thus given rise to a mounting debate about the future role of acaricides in tick and TBD control in southern Africa (Howell et al. 1981; Norval 1981a, 1983b; Ardington 1982; de Vos and Potgieter 1983; Bezuidenhout 1985; Bezuidenhout and Bigalke 1987).

The rationale for rethinking veterinary strategy against ticks and TBDs in southern Africa is partly shared with other parts of the world, as scientific advances make possible new approaches towards animal disease control. However, important elements of the reappraisal of tick and TBD control are unique to southern Africa. Such elements include firstly aspects of tick ecology and TBD epidemiology peculiar to southern Africa and secondly the institutional aspect that a substantial part of the national acaricide usage is funded by the Government in countries such as Zimbabwe.

This paper reviews the circumstances which justify a reappraisal of veterinary strategy towards tick and TBD control in southern Africa, and discusses the scope for new strategies made possible as a result of improved information about tick ecology, TBD epidemiology and the economic impact of tick infestation on livestock productivity. Emphasis is placed on the need for a multidisciplinary approach in the design and appraisal of new veterinary strategies appropriate for southern Africa. As an example of such an approach, the case of Zimbabwe is considered and the scope for moving away from the current strategy of intensive dipping is examined. Finally, broader issues of a social, political and institutional nature are assessed.

2 Arguments Against Intensive Acaricidal Treatment of Cattle

2.1 Acaricide Resistance

The increasingly widespread occurrence of acaricide resistance in southern Africa, particularly for *Boophilus decoloratus* (Baker et al. 1979, 1981; Coetzee et al. 1987) may provide one argument against continuing the intensive acaricidal treatment of cattle. As in eastern Africa (Tatchell 1984), development of resistance in *B. decoloratus* has been the main reason for introduction of new acaricides. Where new acaricides have been introduced to replace the existing ones, these are generally more expensive and it is likely to be only a matter of time before ticks (especially *B. decoloratus*) develop resistance to the new chemicals.

Yet this argument may be partly spurious. *B. decoloratus* had always been considered to be an important pest of cattle, similar to the closely related species *Boophilus microplus* which can cause large production losses and even death in susceptible cattle breeds (Sutherst and Wharton 1971). However, as noted in eastern Africa by Tatchell (1984), the numbers of *B. decoloratus* on cattle are usually too low to cause significant production losses. These observations have been confirmed in studies of tick populations on undipped indigenous cattle (Kaiser et al. 1982, 1988; Pegram et al. 1986; Tatchell and Easton 1986) and it has been shown experimentally that *B. decoloratus* feeds on cattle less successfully than does *B. microplus* (Norval and Short 1984).

Resistance in *B. decoloratus* does not therefore justify changing acaricides in most African situations. Indeed, there are epidemiological reasons why it may be beneficial to tolerate infestations of *B. decoloratus* on cattle. Unlike *B. microplus*, *B. decoloratus* is not a vector of *B. bovis* (Potgieter 1977; de Vos 1979), a parasite which can cause serious disease problems in cattle, but the species does compete with *B. microplus* and excludes this pest from large areas of Africa (Norval and Sutherst 1986; Sutherst 1987). Infestations of *B. decoloratus* are usually indicative of enzootic stability for *B. bigemina* (de Vos and Every 1981; de Vos and Potgieter, 1983; Norval et al. 1983), a situation generally considered to be desirable as it reduces the risk of losses caused by this parasite.

Acaricide resistance will undoubtedly become an increasing problem in the future. However, the key point is that tick infestation per se may have some advantages in certain situations in southern Africa, provided it involves the right tick species and an appropriate epidemiological situation. These potential advantages will depend on environmental conditions and tick resistance of the cattle, both of which influence tick abundance and hence the prevalence of TBDs.

2.2 Production Losses Due to Tick Infestation May Not Justify Acaricide Treatment of Cattle

Although intensive tick control was introduced to southern Africa as a measure to control ECF, veterinarians and farmers subsequently came to believe that the control of ticks per se is of importance in maintaining the health and productivity of cattle. For this reason, intensive acaricide application has continued to be practised in many parts of southern Africa after the eradication of ECF, even where it is not required for the control of other TBDs.

The most abundant and hence important tick pests of cattle in southern Africa are *Rhipicephalus appendiculatus* and *Amblyomma hebraeum* (Theiler

1962; Howell et al. 1978; Norval et al. 1982; Rechav 1982; Norval 1983a). Other species such as *B. microplus* may be very abundant at a local level but are not widely distributed (Howell et al. 1978; Mason and Norval 1980; Baker et al. 1989).

Support for the belief that tick infestations have very large effects on the productivity of cattle was provided by experiments in South Africa by Taylor and Plumb (1981) who showed heavy tick infestations to cause large reductions in liveweight gain (LWG) and death. However, their findings are misleading as they obviously subjected tick-naive cattle to massive challenge. Such challenge and infestations, as noted by Sutherst (1981), do not occur in field situations where calves are born in tick-infested environments and acquire resistance to ticks. Calves normally carry lighter tick burdens than adults (Barnett and Bailey 1955; Sutherst et al. 1979), suggesting that they may be protected by some form of innate, age-related resistance.

In other studies, using previously exposed, resistant cattle it has been shown that production losses caused by *R. appendiculatus* and *A. hebraeum* are not necessarily very large. For example, in Zimbabwe it was found that the larval and nymphal stages caused no detectable losses in LWG (Norval et al. 1988, 1989). The loss in LWG for each adult female that completes engorgement was approximately 4 g in the case of *R. appendiculatus* (Norval et al. 1988) and approximately 10 g in the case of *A. hebraeum* (Norval et al. 1989). Losses in milk production of the order of 7 g for every adult female that completes feeding have been shown to be caused by both species (Norval et al. 1990b, c).

As resistance to these tick species varies between different cattle breeds (Bonsma 1981; Rechav and Zeederberg, 1986; Norval et al. 1988; Spickett 1990) so do the effects of the ticks on the productivity of the different breeds (Norval et al. 1988, 1990d). Tick-susceptible *Bos taurus* breeds become very heavily infested and as a result suffer large losses in both LWG and milk production; with such infestations they also become predisposed to screw-worm fly strike. Indigenous Sanga (*Bos indicus* × *B. taurus*) and Zebu (*B. indicus*) breeds seldom or never become heavily infested and consequently suffer small or insignificant production losses as a result of tick infestation and do not normally become predisposed to secondary screw-worm infestation. The tick resistance of the crosses between *B. taurus* and indigenous breeds is intermediate and so are the effects of tick infestations.

Findings on the effects of ticks on cattle productivity in southern Africa are similar to those made in eastern and central Africa and elsewhere. In Kenya, Tatchell et al. (1986) recorded no differences in LWG between groups of acaricide-treated and untreated cattle in an area that was free of *R. appendiculatus*, and similar results were obtained by Tatchell (1988) in the Sudan.

On the other hand, such findings are not universal. In the Trans-Mara area of Kenya in which *R. appendiculatus* is present, de Castro et al. (1985) recorded higher growth rates in acaricide-treated cattle than in untreated cattle after ECF immunisation and exposure to other TBDs. Similar results were reported by Morzaria et al. (1988) and Mukhebe et al. (1989) from Coast Province in Kenya and in Zambia by Pegram et al. (1991). Estimated losses in LWG per engorged female have been 6 g for *Amblyomma lepidum* (Tatchell 1988) and 46–61 g for *Amblyomma variegatum* (Pegram et al. 1989).

Further data are available from other parts of the world. In Australia, Sutherst et al. (1983) have shown *B. microplus* to cause losses of 0.6–1.5 g per engorged female. In the United States losses per engorged female have been extrapolated by Pegram et al. (1989) as 33 g for *Amblyomma maculatum* (from Williams et al. 1978) and 16 g for *Amblyomma americanum* (from Barnard 1985).

Fewer studies have been carried out elsewhere on the effects of ticks on milk production. In Zambia, Pegram et al. (1991) found that tick infestations, comprising *A. variegatum* and *R. appendiculatus*, caused significant decreases in the amount of milk produced by Sanga cows, and, in the United States, Woodward et al. (1915) demonstrated a significant effect on milk production in dairy cows caused by infestation with *Boophilus annulatus*.

In appraising the need for intensive tick control, the issue is not whether tick infestation leads to any losses in productivity but whether such losses are sufficient to justify expenditure on tick control. There are few published studies of the economics of tick management. Pegram et al. (1991) found significant losses in Sanga cattle caused by mixed tick infestations, but calculated that this did not justify the cost of intensive tick control. However, strategic dipping was demonstrated to be cost-effective. Similar experience has been reported in Australia, where economic analysis showed that strategic acaricidal treatment of Droughtmaster cattle (*B. taurus* × *B. indicus*) infested with *B. microplus* showed substantial benefits in comparison to no treatment (Sing et al. 1983).

The evidence is growing that the production losses caused by ticks in southern Africa are considerably less important than previously believed and may not justify intensive tick control, giving weight to the argument to reappraise veterinary strategy for tick and TBD control in this part of the world. The accuracy of estimates of production losses due to tick infestation is crucial to the economic analysis which is required. These losses will vary according to tick challenge, with the tick species composition being particularly important, and according to the tick resistance of the breeds of cattle present. This has two important implications. Firstly, it is inappropriate to make simple extrapolations to southern Africa from experience in other continents. Secondly, optimal tick control strategy is likely to vary from one situation to another requiring the availability and application of models to extrapolate existing information on tick ecology, TBD epidemiology and veterinary economics to the full range of ecological zones and cattle production systems that exist in southern Africa (Floyd et al. 1987a, b).

2.3 The Financial Burden of Dipping upon Farmers

In an intensive dipping regime designed to achieve total tick control, the acaricide itself is not a large cost, perhaps only US$3 to US$5 per animal per year. For a commercial dairy farmer who sees his animals daily for milking there is little extra handling involved in routine dipping, which therefore does not cost him a great deal relative to the income generated by the cattle. The same situation can apply to intensive beef production. However, in extensive beef production systems, typical of commercial ranches in southern Africa, the rounding up of cattle for dipping is the main activity of ranch labour and often represents the largest single cost of production. Labour is becoming increasingly expensive in southern Africa, where once it was cheap. Commercial beef ranchers will, therefore, welcome a move away from intensive dipping if they can be convinced that alternative strategies are technically sound, cost-effective and do not involve substantial risk of TBD outbreaks.

For peasant farmers whose cattle are dipped at Government dip tanks the true costs of dipping are less obvious. For example, rural farmers in Zimbabwe may be required to take their cattle for compulsory dipping as many as 42 times per year. This results in loss of animal traction and labour for all or part of the days on which dipping occurs. The dip tanks may be as far as 10 km from the farmers' homes, resulting in cattle production losses due to the trekking itself. There are possible losses due to stress-induced abortions, drowning and physical injury during dipping. Where the acaricide cost is recovered from peasant farmers, as in Zimbabwe for a period prior to independence, such fees may represent the largest single financial input to cattle production for such farmers. Rural farmers will welcome the end of compulsory intensive dipping where it is recognised to be unnecessary.

2.4 Constraints on Government Veterinary Programmes in Southern Africa

In several countries in southern Africa, Government veterinary services have a special role in tick and TBD control, being responsible for the funding and implementation of extensive rural dipping programmes for pastoralists and peasant farming communities. These services have features which are different from the public sector veterinary programmes in other continents. In the United States, public funding for compulsory cattle dipping was justified as part of a national eradication campaign. In Australia, compulsory dipping is Government-funded in New South Wales only for the purpose of placing a strategic barrier to the southward expansion of the area which is presently tick-infested. Apart from specific circumstances such as these, where public interest is at stake, tick control in other countries is most commonly the financial responsibility of the farmer himself.

The justification for public sector funding of veterinary services for peasant farmers in southern Africa is a thorny issue. In the case of tick and TBD control, compulsory dipping in southern African countries may at one time have been thought to be leading towards eventual tick eradication, which would have provided a possible case for public funding, but today eradication is an unlikely prospect. There may have been an underlying rationale that commercial farmers benefited by keeping under control the tick populations on peasant-owned cattle herds. The evidence on this point is now recognised to be to the contrary: in Zimbabwe for example much of the grazing land in peasant farming areas is so severely grazed that tick populations are minimal in comparison with neighbouring commercial ranches, where lower stocking rates lead to environments much more favourable to ticks (Norval 1977). One factor contributing to continuing public sector funding may be that free veterinary services provide a way of transferring resources to poor rural farmers.

Zimbabwe is a country where public-sector involvement in rural cattle dipping is substantial. The Government maintains a national network of some 2300 cattle dips serving just under 4 million cattle owned by so-called communal farmers. Total Government expenditure on dipping in 1988/89 amounted to approximately US$9.3 million[5], as broken down in Table 1. The direct costs of cattle dipping in the communal areas constituted 57% of the total budget of the Department of Veterinary Services (excluding expenditure on tsetse and trypanosomiasis control). Without cost recovery from rural farmers, this represents a substantial burden on the national treasury, particularly as the acaricides are purchased externally using foreign exchange which is typically in short supply.

In some southern African countries the ability of Government agencies to carry out effective veterinary programmes is constrained by general infrastructural and institutional problems. Poor public sector salaries have contributed to shortage of sufficient staff with appropriate skills and experience. Financial constraints have meant that field staff often do not have the necessary vehicles, acaricides and drugs. All in all, the practical scope for moving away

Table 1. Department of Veterinary Services, Zimbabwe: breakdown of costs of current Dipping Service, 1988/89 (Perry et al. 1990b; from Dept. of Veterinary Services records)

Item	Expenditure (derived)	
	Amount, Z$ '000	Total (%)
Salaries, bonus and cycle allowance for dip attendants	2816	15.2
Water carrier wages	2108	11.4
Water cart allowance	46	0.3
Dipping chemicals	5210	28.2
Repairs and maintenance of diptanks	1296	7.0
Construction of new diptanks	291	1.6
Diptank record books and supplies	100	0.5
Protected clothing for dip attendants and water carriers	383	2.1
Water pumps	82	0.4
Overhead costs[a]	6157	33.3
Total	18489	100.0

Number of cattle dipped per year: 3.9 million.
Cost of dipping service per head of cattle: ZW$4.74

[a] Overhead costs were assumed to constitute 33.3 % of the total cost of dipping services; they cover costs of Headquarter, Provincial and District staff, depreciation, support facilities, equipment, training and research.

[5] US$1.00 = approximately Z$2.00 in 1988/89.

from intensive dipping of cattle is therefore of urgent interest to Veterinary Departments in southern Africa.

3 Technical Grounds for Considering New Approaches

3.1 Endemic Stability

The possibility of separating tick and TBD control, through new methods of controlling TBDs, allows tick control to be approached with a view to cost-effective management rather than eradication. We face the prospect of being able to live with TBDs either through vaccination or through the achievement of enzootic stability, of which important knowledge has emerged only in the last 15 years (Callow 1977).

It is now known that endemic stability for heartwater, anaplasmosis and babesiosis is a commonly occurring state (de Vos 1979; de Vos and Every 1981; de Vos and Potgieter 1983; Howell et al. 1981; Norval 1981b; Norval et al. 1983, 1984; Bezuidenhout 1985; Bezuidenhout and Bigalke 1987). Endemic stability for *T. parva* also occurs commonly (Norval et al. 1985; Koch et al. 1989).

3.2 Vaccines Against TBDs

The most significant scientific advance allowing us to contemplate new tick and TBD control strategies is the prospect of safe, cheap and effective vaccines becoming available. Blood-based vaccines for babesiosis, anaplasmosis and heartwater have been available for several years (FAO 1984) but their use has been limited because of the problems associated with live vaccines and blood-based products. Considerable progress is now being made in the development of synthetic vaccines against other TBDs using biotechnology (Wright and Riddles 1989). The use of a tick stabilate vaccine for the control of *T. parva* has also been demonstrated (Koch et al. 1989; Irvin et al. 1989).

3.3 Vaccines Against Ticks

There is also the prospect of developing vaccines against ticks: Willadsen et al. (1989) have reported the identification of a protective antigen for *B. microplus* and research on immunological approaches to the control of ticks is being carried out in a number of laboratories around the world. Tick vaccines are likely to be commercially available in Australia in the foreseeable future. It appears to be only a matter of time before equivalent vaccines appropriate to local tick species will be developed for southern Africa. However, the data required to assess the cost effectiveness of tick vaccines versus acaricides for tick control are not yet available.

3.4 Modelling of Tick Ecology, TBD Epidemiology and Veterinary Economics

If TBDs can be controlled by means other than tick control, then the use of acaricides on cattle is economically justified only in relation to the direct effects of the ticks per se on livestock productivity. The principles of veterinary economics to be applied in this type of analysis are straightforward (Morris and Meek 1980; FAO 1983; Putt et al. 1987). The objective becomes to identify a cost-optimal strategic regime of acaricide treatment rather than to achieve total tick control. The practicality of such an economic analysis depends upon our ability to model the impact of tick infestation on livestock productivity for different environments.

Tick abundance on cattle in the absence of control is determined by the susceptibility of the cattle to ticks, the suitability of the environment (climate and vegetation) for tick survival and other factors such as the presence of alternate hosts. In *R. appendiculatus* and *A. hebraeum* the occurrence of the damaging adult stage is also dependent on season: these ticks are most abundant during the summer rains. The consequence of these varying factors is that tick control strategies will vary from place to place. On one hand, susceptible cattle kept in environments that are favourable for tick survival will probably require intensive acaricidal treatment for the greater part of the year. On the other hand, resistant breeds kept in environments marginal to tick survival will probably require little or no tick control.

In order to choose the optimal level of treatment, we must be able to model the impact of different scenarios of intervention upon tick infestation. Our understanding of the complexities of tick population dynamics, breed susceptibility to tick infestation, production losses caused by ticks (damage coefficients) in relation to differing levels of tick control is now approaching the point where we can evaluate alternative tick control strategies with some confidence. In view of the complexity of the interactions between these factors, mathematical models have proved invaluable in estimating quantitative relationships.

For example, a tick population model initially developed for *B. microplus* in Australia (Sutherst and Wharton 1971; Sutherst and Dallwitz 1979) has now been adapted for *R. appendiculatus* in Africa (Floyd et al. 1987a; Sutherst et al. 1987). The model, known as T3HOST, can be used to simulate the effectiveness of control strategies in different environments with different breeds of cattle and varying numbers of alternative hosts (Floyd et al. 1987b). The cost effectiveness of tick control can then be determined using estimates of production losses based on tick numbers, damage coefficients, the market value of cattle products and the current costs of tick control (Floyd et al. 1987b). The costs and benefits of different strategies for the control of *R. appendiculatus* on Sanga cattle at Lake McIlwaine in the highveld of Zimbabwe, estimated by Floyd et al. (1987b) are shown in Fig. 1.

Computer simulations using the T3HOST model have been used to identify the most efficient and economical tick control strategies for Burundi (Kaiser et al. 1988) and these are now being implemented on a large scale in that country (Moran and Nigarura 1990). As the T3HOST model is climate-driven (i.e. actual

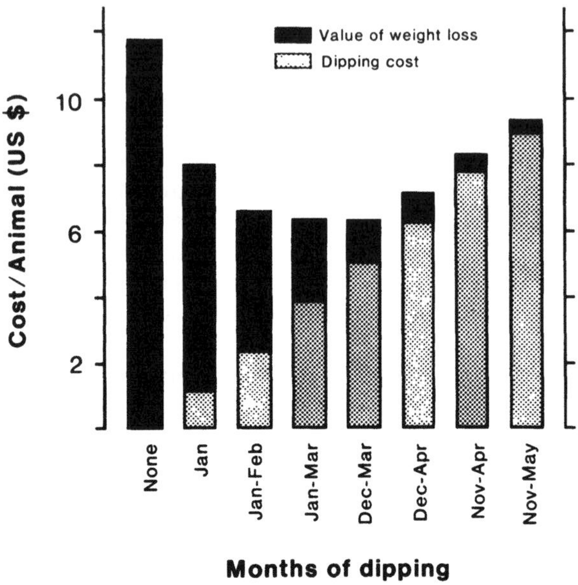

Fig. 1. The costs and benefits of different strategies for the control of *R. appendiculatus* on Sanga cattle at Lake McIlwaine, Zimbabwe

tick data are not required to run simulations) the same approach could be adopted on a much wider scale if satisfactory results are obtained in Burundi.

For southern Africa, a population model for *A. hebraeum* still needs to be developed and the strategies derived from computer simulation need to be field tested.

4 Zimbabwe: the Scope for Moving Away from Intensive Dipping

The design and appraisal of new strategies towards control of ticks and TBDs must take into account our full knowledge of tick ecology and TBD epidemiology. Choice of strategy must be based on rigorous economic analysis of the costs and benefits for each specific ecological zone and farming system. The authors of this paper have recently collaborated to assess current and alternative tick and TBD control strategies in Zimbabwe. Some of the results are presented here in order to illustrate the multidisciplinary approach considered appropriate for planning veterinary strategy towards control of ticks and TBDs in southern Africa.

Regular compulsory dipping of cattle has been in force in Zimbabwe since 1914, initially introduced to control ECF. Following apparant ECF eradication in 1954, compulsory dipping continued, directed against theileriosis, babesiosis, anaplasmosis, heartwater and the effects of ticks per se.

Heavy losses from the TBDs were experienced in the 1970s following a breakdown of dipping services during the independence war. In the 1980s there was a steady resumption of compulsory dipping in the Communal Lands, which is now as intensive in many parts of the country as they have ever been. This has been augmented by a replacement of Delnav (dioxathion) as the main acaricide by Triatix (amitraz), which has considerable residual activity.

The financial and administrative responsibility for tick and TBD control in the Communal Lands is entirely borne by the Government, at an estimated cost of Z$18.5 million (US$9.3 million) for the financial year 1988/89 (Perry et al. 1990b). This is equivalent to an expenditure of Z$4.74 per animal per year (Table 1). As will be shown, this expenditure is rather less than that needed to sustain an intensive dipping strategy. This partly reflects financial and operational constraints upon the Department of Veterinary Services, but also reflects an acceptance by the Department that a move towards less intensive dipping is not only unavoidable but desirable. The following analysis examines the implications of moving in this direction.

A preliminary assessment has been made of alternative TBD control strategies (Perry et al. 1990b) which make less intensive use of expensive acaricides and rely more on controlled immunization and the development of natural immunity to TBDs. The control strategies examined were:

1. intensive dipping: weekly during the summer months and fortnightly for the rest of the year (equivalent of 42 acaricide immersions annually);

Table 2. Criteria for preliminary zonation for tick-borne disease control in Communal Lands in Zimbabwe

Zone[a]	Target diseases	Dipping regimen	Immunization regimen	Approx. no. of cattle ('000)
1.	Babesiosis Anaplasmosis	Minimal	Babesiosis and anaplasmosis in transition	1 872 (48%)
2.	Babesiosis Anaplasmosis Theileriosis	Strategic	Babesiosis and anaplasmosis in transition, theileriosis if necessary	117 (3%)
3.	Babesiosis Anaplasmosis Heartwater	Minimal	Babesiosis, anaplasmosis and heartwater in transition	1 287 (33%)
4.	Babesiosis Anaplasmosis Heartwater	Minimal	Babesiosis and anaplasmosis in transition; heartwater on a regular basis	634 (16%)

[a] The zones described are shown in Fig. 2.

2. reduced dipping – fortnightly in the summer, and monthly for the rest of the year (equivalent to 21 immersions annually); and
3. a combination of strategic (weekly dipping during the summer months, equivalent to 12 immersions, supplemented by natural or artificially induced herd immunity to TBDs) and minimal dipping (equivalent to four acaricide immersions).

The choice of strategic or minimal dipping, and the selection of diseases for immunisation, varies regionally in Zimbabwe, as summarized in Table 2. A provisional division of the Communal Lands into four zones appropriate for the different TBD control regimes is shown in Fig. 2.

For each control strategy, target cattle populations were identified by zone, and projections of the likely consequences were made. It was assumed that there would be no significant differences in disease risk or the effect on livestock production by moving from intensive to either of the alternative strategies. Furthermore, it was also assumed that strategic and minimal dipping would be applied after a 2-year transitional period of reduced dipping.

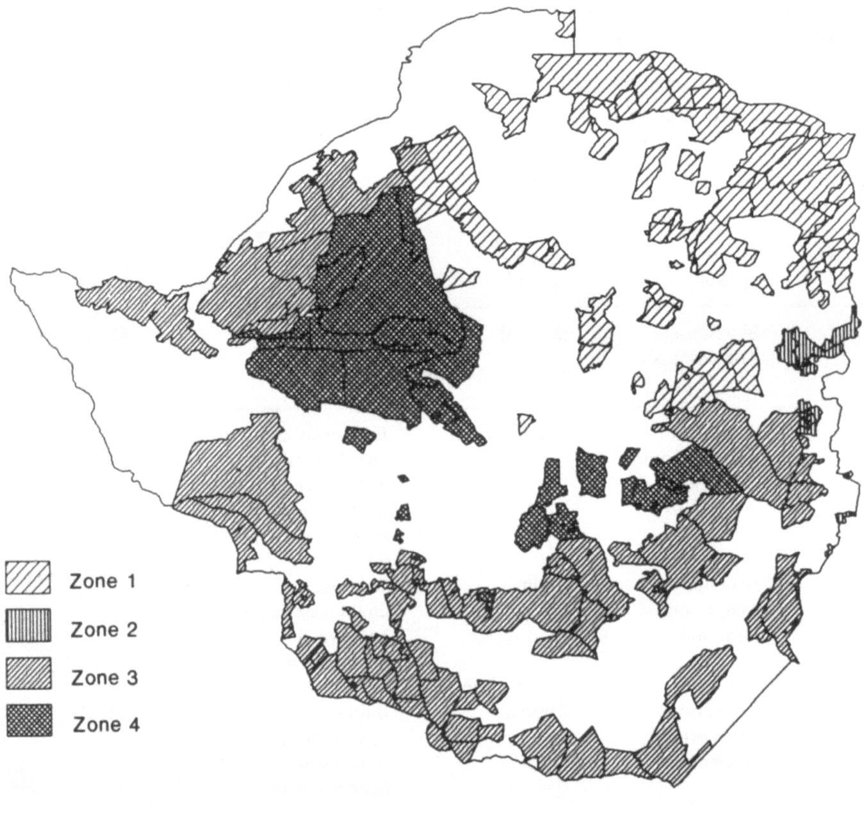

Fig. 2. A preliminary division of Communal Lands in Zimbabwe into zones for tick-borne disease regimes (Perry et al. 1990b)

Table 3. Department of Veterinary Services, Zimbabwe: comparative cost analysis of alternative tick and tick-borne disease control strategies over a 20 year period (Perry et al. 1990b)

	Undiscounted values			Discounted values at					
				10%			14%		
Strategy[a]	A	B	C	A	B	C	A	B	C
Total cost, year 1–20 Z$ '000	1068, 545	557, 104	238, 444	328, 305	176, 973	106, 048	228, 350	125, 175	84, 967
Cost per animal Z$	8.13	4.24	1.81	2.50	1.35	0.81	1.74	0.95	0.65

[a] A, Intensive dipping strategy; B, reduced dipping strategy; C, strategic and minimal dipping strategy.

Resources (acaricides, personnel costs, facilities, supplies and overheads) expended under each strategy were costed and projected over a 20-year period. A summary of the projected costs for the three strategies, discounted at 10 and 14%, is provided in Table 3. Current annual expenditure of Z$4.74 per animal (Table 1) is closer to the estimated cost of Z$4.24 for reduced dipping (undiscounted cost for strategy B, Table 3) than the cost of Z$8.13 for intensive dipping. While the recent policy of the Department has been to reduce the intensity of dipping throughout the country, the current average expenditure also reflects a combination of relatively intensive dipping in many areas of the highveld and little dipping in areas of the lowveld, for example where water supply is problematic.

Over a 20-year period, both the reduced and strategic-minimal dipping strategies would be more cost-effective than intensive dipping (ZW$2.50 per animal per year) by 46% (ZW$1.35) for reduced dipping and by 68% (ZW$0.81) for strategic-minimal dipping (in present value terms, at a 10% discount rate). Furthermore, these figures also clearly demonstrate that the strategic-minimal dipping strategy would be more cost-effective than the reduced dipping strategy, and both options are likely to result in cost savings over present dipping practice.

This preliminary assessment necessarily made crucial assumptions that require to be tested thoroughly before the Veterinary Department in Zimbabwe can commit itself to a change in strategy. However, the magnitude of the projected economic benefits provides strong justification for ongoing technical and socio-economic evaluation of the alternative strategies which have been described. Such investigations are being pursued in a collaborative study involving the International Laboratory for Research in Animal Diseases, the University of Florida and staff of the Department of Veterinary Services in Harare.

5 Discussion

5.1 Financial and Economic Analysis

While there still remains the need to obtain more data on tick ecology, TBD epidemiology and livestock production losses under varying situations of exposure to ticks and TBDs in southern Africa, much information is already available. An important requirement now is the development of epidemiological models which can be used to predict the effect on TBD of tick control strategies in the various ecological zones. This scientific research must be complemented by and oriented towards an economic framework for the design and appraisal of veterinary strategy.

The importance of being able to quantify the production losses associated with different levels of tick infestation was discussed in Section 2.2. Subsequently there arises the problem of valuing such losses. This will depend upon the type of production system involved, which varies considerably throughout southern Africa. Where cattle are reared principally for commercial beef production, be it intensive commercial farming or extensive pastoral systems, the valuation of potential production losses due to tick infestation can be estimated from meat prices in domestic or international markets where appropriate.

Greater problems of analysis arise in those parts of southern Africa where livestock are not primarily kept for beef production but rather as part of a traditional agropastoral farming system where their primary output is animal draught. In Zimbabwe for example, this is precisely the part of the country where Government-funded cattle dipping takes place. This means that careful attention must be paid to the economic role of cattle in the farming system, particularly the valuation of draught output (Barrett 1991), if economic analysis of alternative veterinary strategies is to be comprehensive and valid.

Financial and economic analysis should also address the indirect costs and benefits associated with changes in veterinary strategy. For peasant farmers, indirect costs will result from trekking animals for compulsory acaricide treatment at Government dip tanks, including loss of animal draught, labour, and other costs discussed in Section 2.3. There may also be social costs associated with intensive dipping; for example, cattle trekking is known to be a common cause of gully erosion, contributing to environmental degradation.

Changes of Government policy towards tick and TBD control in areas of peasant agriculture and livestock production may have consequences for commercial cattle producers. Economic analysis of changes in veterinary strategy must take into account such linkages, the implications of which are not always clear. On the one hand, commercial farmers might tend to shift from production of exotic tick-susceptible breeds into indigenous breeds and reduce their acaricide usage. On the other hand, there may be stepping up of acaricide use if farmers continue with exotic breeds and perceive an increase in disease risk or are uncertain about the effectiveness of new veterinary strategies being adopted by the Government.

5.2 Epidemiological Zoning

At present it is possible to define the broad epidemiological zones for each of the major diseases only on fairly subjective grounds (Howell et al. 1981; Norval 1981a). The effective management and economic analysis of TBD control in southern Africa will be greatly improved if we can define ecological/epidemiological zones more objectively. As a first step in this direction, geographic information systems are proving promising in the investigation of the relationship between climate and vegetation and the distribution of *R. appendiculatus* in Africa (Kruska and Perry 1990; Perry et al. 1990a; 1991) as well as factors of importance in the epidemiology of *T. parva* (Lessard et al. 1990).

5.3 Sero-Diagnostic Requirements

In order to assess the scope for reduced or minimal dipping strategies in a specific area it is important to be able to assess the prevalence of antibodies to

TBDs in that area as an estimate of the immune status of the animal population. With diseases such as heartwater and theileriosis it may also be necessary to differentiate between immunologically different strains. At present, the sero-diagnostic methods and facilities appropriate for this type of survey are not available in all of the countries of southern Africa.

Zonation of countries into areas where different veterinary strategies are applied could cause problems associated with livestock movement. For example, animals in an area of minimal dipping are likely to acquire a carrier status for some TBDs. If these animals were subsequently transferred to an area of intensive tick control where cattle had not been exposed to the diseases, outbreaks could occur. While vaccination would provide a partial solution to the problem, it will be necessary to have appropriate sero-diagnostic facilities so that farmers can be advised of the risks associated with such movements. The costs of establishing and maintaining such a diagnostic service will have to be taken into account in appraising the economics of tick control strategies based on reduced use of acaricides.

5.4 Social, Institutional and Political Considerations

The appraisal of new strategies must go beyond technical and economic feasibility to consider also social, institutional and political aspects of major changes in Government policy towards tick and TBD control in rural areas. For example, in Zimbabwe weekly or fortnightly attendance at the cattle dip is an important forum for interface between cattle owners and Government veterinary officials, where cattle statistics are collected, cattle movement permits are authorised and cattle may be inspected, vaccinated or otherwise treated. Because of the widespread and substantial number of cattle deaths which occurred in the communal areas of Zimbabwe where veterinary services were suspended during the independence war, rural farmers are mostly very keen to have their cattle dipped; it could be an unwelcome political issue if it were perceived that the Government was simply trying to reduce its expenditure on veterinary services in the rural areas.

While it is currently feasible to produce blood-based vaccines for anaplasmosis, babesiosis and heartwater, there are obvious risks, involving adequate attenuation of strains and the presence of other blood-borne pathogens, associated with the use of such vaccines within and between African countries. Prerequisites for the production of blood-based vaccines are good laboratory facilities and access to disease-free herds. Government Veterinary Departments are naturally reluctant to make changes in technical policy which place them at high risk if a sole external source of essential vaccines may prove unreliable at some time in the future.

5.6 Quo Vadis

While such issues are often not of great interest to the scientific research worker, they must be squarely addressed if new veterinary strategies are to be transferred from the academic literature to be applied in the field. Government planners and senior staff in the Veterinary Departments of southern Africa require to be convinced not only that the current strategies are inappropriate, but also that proposed new strategies are sound in every respect: technical, financial, economic, social, political and institutional. With an appropriate multidisciplinary approach, this may be a real prospect.

Acknowledgements. We thank the Director of Veterinary Services of Zimbabwe for provision of and permission to publish detailed information on the costs of tick control in the Communal Lands. This study was funded in part by the International Laboratory for Research on Animal Diseases, and the United States Agency for International Development (Contract No. AFR-0435-A-00-9084-00) through the University of Florida. At the time of the study John Barrett was attached to the Department of Veterinary Services in Harare as a Technical Cooperation Officer funded by the British Overseas Development Administration.

References

Ardington PC (1982) The benefits of intensive tick control. In: Proc Symp Ectoparasites of cattle. South African Bureau of Standards, Pretoria, S Africa, pp 136–140

Baker JAF, Jordaan JO, Robertson WD (1979) Ixodicidal resistance in *Boophilus microplus* (Canestrini) in the Republic of South Africa and Transkei. J S Afr Vet Assoc 50:296–310

Baker JAF, Jordaan JO, Robertson WD (1981) A comparison of the resistance spectra to ixodicides of *Boophilus decoloratus* (Koch) and *Boophilus microplus* (Canestrini) in the Republic of South Africa and Transkei. In: Whitehead GB, Gibson JD (eds) Tick biology and control. Tick Research Unit, Rhodes University, Grahamstown, pp 103–108

Baker MK, Ducasse FBW, Sutherst RW, Maywald GF (1989) The seasonal tick populations on traditional and commercial cattle grazed at four altitudes in Natal. J S Afr Vet Assoc 60:95–101

Barnard DR (1985) Injury thresholds and production loss functions for the Lone Star Tick *Amblyomma americanum* (Acari: Ixodidae) on pastured, preweaner beef cattle, *Bos taurus*. J Econ Entomol 78:852–855

Barnett SF, Bailey KP (1955) A field survey of East Coast fever and calf mortality in an enzootic E.C.F. area. East African Veterinary Organization Annual Report 1954–1955, Nairobi, pp 51–62

Barrett JC (1991) Valuing animal draught in agropastoral farming systems in Zimbabwe. Proc Worksh Tillage, past and future, held at the Institute of Agricultural Engineering, Harare, Zimbabwe Nov 14, 1989. CIMMYT Regional Office, P.O. Box MP 154, Mount Pleasant, Harare, Zimbabwe. CIMMYT, FSR Net Worksh Rep No 22, pp 21–30

Bezuidenhout JD (1985) The epizootiology and control of heartwater and other tick-borne diseases of cattle in South Africa. Onderstepoort J Vet Res 52:211–214

Bezuidenhout JD, Bigalke RB (1987) The control of heartwater by means of tick control. Onderstepoort J Vet Res 54:525–528

Bonsma J (1981) Breeding tick-repellent cattle. In: Whitehead GB, Gibson JD (eds) Tick biology and control. Tick Research Unit, Rhodes University, Grahamstown, pp 67–77

Callow LL (1977) Vaccination against bovine babesiosis. In: Miller LH, Pino JA, McKelvey JJ (eds) Immunity to blood parasites of man and animals. Plenum, New York, 121 pp

Coetzee BB, Stanford GD, Davis DAT (1987) The resistance spectrum shown by a fenvalerate-resistant strain of blue tick (*Boophilus decoloratus*) to a range of ixodicides. Onderstepoort J Vet Res 54:79–82

de Castro JJ, Young AS, Dransfield RD, Cunningham MP, Dolan TT (1985) Effects of tick infestation on Boran (*Bos indicus*) cattle immunized against theileriosis in an endemic area of Kenya. Res Vet Sci 39:279–288

de Vos AJ (1979) Epidemiology and control of bovine babesiosis in South Africa. J S Afr Vet Assoc 50:357–362

de Vos AJ, Every R (1981) Epidemiology and control of bovine babesiosis in the Natal midlands. In: Whitehead GB, Gibson JD (eds) Tick biology and control. Tick Research Unit, Rhodes University, Grahamstown, pp 57–60

de Vos AJ, Potgieter FT (1983) The effect of tick control on the epidemiology of bovine babesiosis. Onderstepoort J Vet Res 50:3–5
FAO (1983) Report of the FAO expert consultation on the economics of tick control, Gainesville, Florida, 16–18 May 1983. Food and Agriculture Organisation of the United Nations, Rome
FAO (1984) Ticks and tick-borne disease control. A practical field manual, vol II. Tick-borne disease control. Food and Agriculture Organisation of the United Nations, Rome, 320 p
Floyd RB, Maywald GF, Sutherst RW (1987a) Ecological models. 2. A population model of *Rhipicephalus appendiculatus*. In: Sutherst RW (ed) Ticks and tick-borne diseases. ACIAR Proc No 17, Canberra, pp 72–75
Floyd RB, Sutherst RW, Maywald GF (1987b) Modelling tick management. 3. Designing control strategies for *Rhipicephalus appendiculatus* using T3HOST. In: Sutherst RW (ed) Ticks and tick-borne diseases. ACIAR Proc No 17, Canberra, pp 72–75
Howell CJ, Walker JB, Nevill EM (1978) Ticks, mites and insects infesting domestic animals in South Africa. Part 1. Descriptions and biology. Sci Bull No 393, Dept Agric Tech Serv, South Africa, 69 pp
Howell CJ, de Vos AJ, Bezuidenhout JD, Potgieter FT, Barrowman PR (1981) The role of chemical tick eradication in the control or prevention of tick-transmitted diseases of cattle. In: Whitehead GB, Gibson JD (eds) Tick biology and control. Tick Research Unit, Rhodes University, Grahamstown, pp 61–66
Irvin AD, Morzaria SP, Munatswa FC, Norval RAI (1989) Immunization of cattle with a *Theileria parva bovis* stock from Zimbabwe protects against challenge with virulent *T. p. parva* and *T. p. lawrencei* stocks from Kenya. Vet Parasitol 32:271–278
Kaiser MN, Sutherst RW, Bourne AS (1982) The relationship between ticks and zebu cattle in southern Uganda. Trop Anim Health Prod 14:63–74
Kaiser MN, Sutherst RW, Bourne AS, Gorissen L, Floyd RB (1988) Population dynamics of ticks on Ankole cattle in five ecological zones in Burundi and strategies for their control. Prev Vet Med 6:199–222
Koch HT, Kavemba L, Ocama JGR, Munatswa FC, Dolan TT, Norval RAI (1989) Immunization of cattle against *Theileria parva bovis* in Zimbabwe. In: Dolan TT (ed) Theileriosis in eastern, central and southern Africa. Proc Worksh on East Coast fever immunization, held in Lilongwe, Malawi, 20–22 Sept 1988. ILRAD, Nairobi pp 90–92
Kruska RL, Perry BD (1990) Evaluation of grazing lands of Zimbabwe using AVHRR normalized difference vegetation index. Prev Vet Med 11:361–363
Lawrence JA (1992) History of bovine theileriosis in southern Africa. In: Norval RAI, Perry BD, Young AS. The epidemiology of theileriosis in Africa. Academic Press, London, pp 1–39
Lawrence JA, Foggin CM, Norval RAI (1980) The effects of war on the control of diseases of livestock in Rhodesia (Zimbabwe). Vet Rec 107:82–85
Lessard P, L'Eplattenier R, Norval RAI, Kundert K, Dolan TT, Croze H, Walker JB, Irvin AD, Perry BD (1990) Geographical information systems for studying the epidemiology of cattle diseases caused by *Theileria parva*. Vet Rec 126:255–262
Mason CA, Norval RAI (1980) The ticks of Zimbabwe. 1. The genus *Boophilus*. Zimbabwe Vet J 11:36–43
Moran MC, Nigarura G (1990) Strategic tick control in Burundi. Parasitologia 32:177–184
Morris RS, Meek AH (1980) Measurement and evaluation of the economic effects of parasitic disease. Vet Parasitol 6:165–184
Morzaria SP, Irvin AD, Waithanga J, D'Souza D, Katende J, Young AS, Scott J, Gettinby G (1988) The effect of East Coast fever immunization and different acaricide treatments on productivity of beef cattle. Vet Rec 123:313–320
Mukhebi AW, Wathanga J, Perry BD, Irvin AD, Mozaria SP (1989) Financial analysis of East Coast fever control strategies on beef production under farm conditions. Vet Rec 125:456–459
Norval RAI (1977) Tick problems in relation to land utilisation in Rhodesia. Rhod Vet J 8:33–38
Norval RAI (1981a) A re-assessment of the role of dipping in the control of tick-borne disease in Zimbabwe. In: Whitehead GB, Gibson JD (eds) Tick biology and control. Tick Research Unit, Rhodes University, Grahamstown, pp 87–90
Norval RAI (1981b) Heartwater in Tribal Trust Lands in southern Zimbabwe. Zimbabwe Vet J 12:56–57

Norval RAI (1983a) The ticks of Zimbabwe. VII. The genus *Amblyomma*. Zimbabwe Vet J 14:3–18

Norval RAI (1983b) Arguments against intensive dipping. Zimbabwe Vet J 14:19–25

Norval RAI, Short NJ (1984) Interspecific competition between *Boophilus decoloratus* and *Boophilus microplus* in southern Africa. In: Griffiths DA, Bowman CE (eds) Acarology VI. vol. 2. Ellis Horwood, Chichester, pp 1242–1246

Norval RAI, Sutherst RW (1986) Assortative mating between *Boophilus decoloratus* and *Boophilus microplus* (Acari: Ixodidae). J Med Entomol 23:459–460

Norval RAI, Walker JB, Colborne J (1982) The ecology of *Rhipicephalus zambeziensis* and *Rhipicephalus appendiculatus* (Acarina, Ixodidae) with particular reference to Zimbabwe. Onderstepoort J Vet Res 49:181–190

Norval RAI, Fivaz BH, Lawrence JA, Daillecourt T (1983) Epidemiology of tick-borne diseases of cattle in Zimbabwe. 1. Babesiosis. Trop Anim Health Prod 15:87–94

Norval RAI, Fivaz BH, Lawrence JA, Brown AF (1984) Epidemiology of tick-borne diseases of cattle in Zimbabwe. II. Anaplasmosis. Trop Anim Health Prod 16:63–70

Norval RAI, Fivaz BH, Lawrence JA, Brown AF (1985) Epidemiology of tick-borne diseases of cattle in Zimbabwe. III. The *Theileria parva* group. Trop Anim Health Prod 17:19–28

Norval RAI, Sutherst RW, Kurki J, Gibson JD, Kerr JD (1988) The effect of the brown ear-tick *Rhipicephalus appendiculatus* on the growth of Sanga and European breed cattle. Vet Parasitol 30:149–164

Norval RAI, Sutherst RW, Jorgensen OG, Gibson JD, Kerr JD (1989) The effect of the bont tick (*Amblyomma hebraeum*) on the weight gain of Africander steers. Vet Parasitol 33:329–341

Norval RAI, Lawrence JA, Young AS, Perry BD, Dolan TT, Scott J (1991) *Theileria parva*: Influence of vector, parasite and host relationships on the nature and distribution of theileriosis in southern Africa. Parasitology 102:347–356

Norval RAI, Sutherst RW, Kurki J, Kerr JD, Gibson JD (1990b) The effects of the brown ear-tick *Rhipicephalus appendiculatus* on milk production of Sanga cattle (in preparation)

Norval RAI, Sutherst RW, Jorgensen OG, Kerr JD (1990c) The effects of the bont tick (*Amblyomma hebraeum*) on milk production of Sanga and Sanga × Brahman cattle (in preparation)

Norval RAI, Sutherst RW, Kerr JD (1990d) The resistance of different breeds of cattle to the bont tick (*Amblyomma hebraeum*) (in preparation)

Norval RAI, Perry BD, Young AS (1992) The epidemiology of theileriosis in Africa, Academic Press, London, 481 pp

Pegram RG, Perry BD, Musisi FL, Mwanaumo B (1986) Ecology and phenology of ticks in Zambia: seasonal dynamics on cattle. Appl Exp Acarol 2:25–45

Pegram RG, Lemche J, Chizyuka HGB, Sutherst RW, Floyd RB, Kerr JD, McCosker PJ (1989) Effect of tick control on liveweight gain of cattle in central Zambia. Med Vet Entomol 3:313–320

Pegram RG, James AD, Oosterwijk GPM, Killorn KJ, Lemche J, Ghirotti M, Tekle Z, Chizyuka HGB, Mwase ET, Chizhuka F (1991) Studies on economic impact of ticks in Zambia. Exp Appl Ecol 12:9–26

Perry BD, Lessard P, Norval RAI, Kundert K, Kruska R (1990a) Climate, vegetation and the distribution of *Rhipicephalus appendiculatus* in Africa. Parasitol Today 6:100–104

Perry BD, Mukhebi AW, Norval RAI, Barrett JC (1990b) A preliminary assessment of current and alternative tick and tick-borne disease control strategies in Zimbabwe. Report to the Director of Veterinary Services, Harare (limited distribution)

Perry BD, Kruska RL, Lessard P, Norval RAI, Kundert K (1991) Estimating the distribution and abundance of *Rhipicephalus appendiculatus* in Africa. Prev Vet Med 11:261–268

Potgieter FT (1977) The life cycle of *Babesia bovis* and *Babesia bigemina* in ticks and in cattle in South Africa. PhD Thesis, Rand Afrikaans University, Johannesburg, South Africa

Putt SNH, Shaw APM, Woods AJ, Tyler L, James AD (1987) Veterinary epidemiology and economics in Africa: a manual for use in the design and appraisal of livestock health policy. ILCA Manual No 3, International Livestock Centre for Africa, Addis Ababa, Ethiopia

Rechav Y (1982) Dynamics of tick populations (Acari: Ixodidae) in the Eastern Cape Province of South Africa. J Med Entomol 6:679–700

Rechav Y, Zeederberg ME (1986) Tick populations of two breeds of cattle under field conditions, with a note on blood components related to host resistance. In: Sauer JR, Hair JA (eds)

Morphology, physiology, and behavioral biology of ticks. Ellis Horwood, Chichester, pp 445–456
Sing NC, Johnston LAY, Leatch G (1983) The economics of cattle tick control in dry tropical Australia. Aust Vet J 60:37–39
Spickett AM (1990) Breed and genetic resistance of animals to vectors and disease. In: Coetzer JAW, Kriek NPJ, Thompson GR, Tustin RC (eds) Infectious diseases of livestock in southern Africa. Oxford University Press, Cape Town (in press)
Sutherst RW (1981) Is the Australian pest management approach to tick control relevant to Africa? In: Whitehead GB, Gibson JD (eds) Tick biology and control. Tick Research Unit, Rhodes University, Grahamstown, pp 79–85
Sutherst RW (1987) The dynamics of hybrid zones between tick (Acari) species. Int J Parasitol 17:921–926
Sutherst RW, Dallwitz MJ (1979) Progress in the development of a population model for the cattle tick *Boophilus microplus*. In: Piffe E (ed) Proc 4th Int Congr Acarol, Saaefeldern, Austria, pp 557–563
Sutherst RW, Wharton RH (1971) Preliminary considerations for a population model for *Boophilus microplus* in Australia. In: Proc 3rd Int Congr Acarol, pp 797–801
Sutherst RW, Wharton RH, Cook IM, Sutherland ID, Bourne AS (1979) Long-term population studies on the cattle tick (*Boophilus microplus*) on untreated cattle selected for different levels of tick resistance. Aust J Agric Res 30:353–368
Sutherst RW, Maywald GF, Kerr JD, Stegeman DA (1983) The effect of cattle tick *Boophilus microplus* on the growth of *Bos indicus* × *B. taurus* steers. Aust J Agric Res 34:317–327
Sutherst RW, Floyd RB, Maywald GF, Dallwitz MJ (1987) Modelling tick populations. 1. Introduction. In: Sutherst RW (ed) Ticks and tick-borne diseases. ACIAR Proc No 17, Canberra, pp 36–38
Tatchell RJ (1984) Strategies for tick control following East Coast fever immunization programmes. In: Irvin AD (ed) Immunization against theileriosis in Africa. ILRAD, Nairobi, pp 110–113
Tatchell RJ (1988) A study of the effect of tick infestation on liveweight gain of cattle in the Sudan. Trop Pest Manage 34:165–167
Tatchell RJ, Easton E (1986) Tick (Acari: Ixodidae) ecological studies in Tanzania. Bull Entomol Res 76:229–246
Tatchell RJ, Chimwani D, Chirchir ST, Ongare JO, Mwangi E, Rinkanya F, Whittington D (1986) A study of the justification for intensive tick control in Kenyan rangelands. Vet Rec 119:401–403
Taylor RJ, Plumb IR (1981) The effect of natural tick infestation on various blood components and livemass in the bovine in South Africa. In: Whitehead GB, Gibson JD (eds) Tick biology and control. Tick Research Unit, Rhodes University, Grahamstown, pp 21–28
Theiler G (1962) The Ixodoidea parasites of vertebrates in Africa south of the Sahara (Ethiopian Region). Report to the Director of Veterinary Services, Onderstepoort (S. 9958), 260 pp
Willadsen P, Riding GA, McKenna RV, Kemp DH, Tellam RL, Nielsen JN, Lahnstein J, Cobon GS, Gouch JM (1989) Immunological control of a parasitic arthropod. Identification of a protective antigen from *Boophilus microplus* J Immunol 143:1346–1351
Williams RE, Hair JA, McNew RW (1978) Effects of Gulf Coast ticks on blood composition and weights of pastured Hereford steers. J Parasitol 64:336–342
Woodward TE, Turner WF, Curtice C (1915) The effect of the cattle tick upon the milk production of dairy cows. Bull No 147 US Dept Agric Washington, DC, 22 pp
Wright IG, Riddles PW (1989) Biotechnology in tick-borne diseases: present status, future perspectives. In: Biotechnology for livestock production. FAO and Plenum, New York, pp 325–340

Bio-Economic Impact of *Amblyomma americanum* in Beef Cattle Production Systems

D.R. Barnard[1], R.T. Ervin[2] and F.M. Epplin[3,4]

Summary

Amblyomma americanum (Linnaeus) (Acari: Ixodidae) is indigenous to North America. On beef cattle, these ticks suck blood, damage hides, and cause "tick worry". Since beef production in the USA is a commodity based industry, each factor can affect producer profits.

Integrated pest management (IPM) offers a feasible decision-making framework for tick control on beef cattle. IPM is based on experimentally defined economic injury levels and requires the application of experience, empiricism, and modelling to determine if beef production and pasture management practices can be changed to comply with recommendations for tick control. A damage model fitted to data for daily rates of weight gain in tick infested nursing *Bos taurus* cattle shows that a functional relationship exists between calf growth rates and tick density. Based on this damage model, economic injury levels are 26–38 feeding female ticks per calf when acaricides are used for tick control. Economic injury levels vary with the cost and effectiveness of different acaricides and with the cost of labor for gathering livestock for treatment. The latter factor changes with the size and condition of the pasture. Economic injury levels for non-chemical control of *A. americanum* do not exist because we are unable to predict market value and management cost factors for these technologies relative to improved yield in beef cattle.

Bos indicus cattle may present an alternative to acaricides for tick control because *Bos indicus* cattle are not significantly compromised by tick infestation. Using a mathematical programming model, we found that a *Bos indicus* beef production system generated 19% more return per hectare on tick infested pasture than did a *Bos taurus* beef production system. The incorporation of *Bos indicus* genotypes into beef cattle breeding programs in the USA could raise the economic injury level for *A. americanum*.

1 Introduction

Amblyomma americanum (Linnaeus) (Acari: Ixodidae), the "lone star tick", is indigenous to North America (Cooley and Kohls 1944). Its range extends south and east from central Texas, Oklahoma and southern Missouri to the Gulf of Mexico and the Atlantic coast. Within this area, *A. americanum* shares its habitat with nearly 12 million cows, nursing calves, and stocker beef cattle (USDA 1988). Cattle are readily exploited as a food source by *A. americanum.*

---

[1] Medical and Veterinary Entomology Research Laboratory, ARS, USDA, Gainesville, Florida 32604, USA

[2] Department of Agricultural Economics, Texas Tech University, Lubbock, Texas 79409, USA

[3] Department of Agricultural Economics, Oklahoma State University, Stillwater, Oklahoma 74078, USA

Ultimately, this results in high densities of ticks on pastures (Hair and Howell 1971; Barnard 1981; Barnard et al. 1989).

Ticks suck blood, inject toxin, cause paralysis, damage hides, and cause "tick worry" (Drummond et al. 1988). Precise values for losses caused by *A. americanum* feeding on beef cattle do not exist but annual weight losses in US cattle are estimated at 4.1 kg liveweight per animal (Drummond et al. 1981). At a market value of 1.54 US$/kg, and with about 12 million beef cattle on tick infested pastures, the value of lost gain may be as high as 72.6 million US$ annually. Most of these losses can be attributed to *A. americanum*, a tick that is present in high densities on pastures. It also has a lengthy period of seasonal activity and each active life stage readily feeds on cattle (Drummond 1967; Barnard 1981).

Immediate reductions of *A. americanum* populations on cattle can be achieved by topical applications of acaricide (Drummond et al. 1988). This technique protects cattle from tick infestation for 72 h or less (Barnard and Jones 1981) and remains the primary defense against *A. americanum* in most cow-calf and stocker cattle programs in the US (Drummond et al. 1988). Other techniques for tick control, including pasture rotation (Wilkinson 1957), host exclusion (Bloemer et al. 1986), breeding cattle for resistance to ticks (Strother et al. 1974; Utech et al. 1978; Sutherst and Utech 1981; George et al. 1985; Rechav and Zeederberg 1986; Sutherst et al. 1988) and cattle management schemes that alter tick-host contact or reduce survivorship in fed ticks (Barnard 1988) have been less thoroughly studied. The same is true for vegetation management methods of tick control including mechanical clearing (Clymer et al. 1970; Meyer et al. 1982), burning (Hoch et al. 1971; Barnard 1985b), and the use of herbicides (Wilkinson 1977; Barnard 1985b).

Beef production is a commodity based process. The management of tick populations on beef cattle directly impacts profit. However, the relationship between the cost of treatment for tick control and the effectiveness of the treatment is poorly understood. Even less well known is the extent to which each method of tick control influences weight gain in cattle. Without this information, however, we have no rational basis for managing *A. americanum* populations on beef cattle.

The concept of integrated pest management (IPM) (Stern et al. 1959; Smith and van den Bosch 1967) provides a useful decision-making framework for tick control. In the present case, IPM involves the consolidation of tick control techniques into a unified program, the object of which is to avoid economic and environmental damage while bringing about tick control.

There is a high probability of success using IPM against *A. americanum* because the beef cattle industry in the US is profit oriented. Cattle are raised for fast growth and marketability and they are subject to tick control. Moreover, an absence of significant tick-borne disease in beef cattle lowers the urgency of need for immediate tick control. A low urgency of control facilitates the effective and economical use of long term tick control methods such as animal management and vegetation management.

The focal point for decision-making in IPM is the economic injury level (Stern et al. 1959; Stone and Pedigo 1972). *The economic injury level is the density of parasitic ticks at which the cost of tick control is equivalent to the value of increased liveweight gain.* When tick densities are below the economic injury level and cattle are treated for ticks, the value of the cattle is reduced by the cost of the control measure. When tick densities exceed the economic injury level, the application of control measures increases the value of the cattle.

Cost-effectiveness ratios are known for some acaricides (Barnard and Jones 1981). These ratios permit identification of the economic injury level which, in turn, provide a basis for defining economic loss in beef cattle caused by tick feeding. In relation to the whole farm enterprise, economic injury levels are the critical component in analyses of tick management strategies for beef cattle (Ervin et al. 1985).

Before holistic management strategies for *A. americanum* on beef cattle can be formulated, estimates of relevant parameters are needed. Two key issues that can be examined experimentally are:

1. Does *A. americanum* reduce liveweight gain in beef cattle?
2. What are the economic injury levels for chemical and alternative control methods against *A. americanum* on beef cattle?

Thereafter, a blend of practical experience, empiricism, and mathematical modelling is needed to determine if beef production and pasture management practices can be responsive to recommendations for effective and economical tick control. We must also consider if the resources required to accomplish tick control can be used more efficiently for other purposes within the whole farm framework.

The eventual success of IPM depends on user (producer) acceptance. However, user acceptance is the least measurable and the least predictable parameter affecting IPM.

In the remainder of this chapter we detail the state of knowledge pertaining to the biologic and economic impact of *A. americanum* on beef cattle. To date, in the US, whole systems management strategies have not been directed at *A. americanum* on beef cattle. Conventional efforts to manage these ticks continue but these efforts are driven by a fragmentary understanding of economic, environmental, beef production, and control efficacy variables. The main constraint to efficient management of *A. americanum* is the limited knowledge of key parameters and a limited research effort to devise, test, and implement holistic management strategies.

Finally, we present one example of how a mathematical programming approach may be used to investigate the sensitivity of estimates of various beef production parameters. The results suggest that losses in beef cattle caused by *A. americanum* can be reduced by selection of alternative breeds of cattle.

2 Biological Impact of *Amblyomma americanum* Populations on Preweaner Beef Cattle

Poor growth is symptomatic of tick infestation in cattle. Moreover, we suppose a functional relationship to exist between tick density and calf growth rate. Based on this supposition, we sought to characterize tick damage to preweaner *Bos taurus* cattle in a study in Oklahoma, USA (Barnard 1985a). Average daily gain (ADG) in nursing calves infested with *A. americanum* is compared to ADG in calves that are not tick infested. The comparison was made for a 2-year period (1982–1983) and comprised eleven separate trials.

Identifiable sources of systematic and experimental error were minimized before the study. The pasture was 25 ha and was free of *A. americanum.* We used calves of comparable age and genotype. Estrous cycles in 15 *B. taurus* cows were synchronized using dinoprost tromethamine in 1980. The cows were artificially inseminated with semen from one *B. taurus* sire and the calves were born each year in February and March.

Cows were treated for endoparasites in May and October, using levamisole hydrochloride per os, and for *Hypoderma* spp. in June using famphur. Also in June, fenvalerate ear tags were applied to selected cows to keep densities of *Haematobia irritans* $\leq$ 25 per animal.

Calf weights were obtained using a conventional weighing regimen. Cows and calves were separated 2 h after sunrise and the calves were held in shaded pens without food or water. After 8 h, each calf was weighed and then released. Average daily gain (ADG) was calculated to the nearest 0.01 kg (Hughes 1976). We studied only calves whose ADG fit the interval $\mathrm{MEAN}_{\mathrm{ADG}} \pm 1$ standard error of the mean (SEM). The interval was defined by measuring ADG for each calf every 7 days during a 4–6 week period which began 2 weeks after the mean birth date of all calves. Calves chosen were placed at random in either a treatment (tick-infested) or control (tick-free) group. Male and female calves were about evenly divided among the groups in both years ($\mathrm{n}_{\mathrm{treatment}} = 6$; $\mathrm{n}_{\mathrm{control}} = 5$).

Five trials were undertaken in 1982 and six in 1983. In each trial, which lasted 28 days, ADG in treatment calves was compared to ADG in control calves. On the first day in each trial, all calves were weighed and the treatment calves infested with 250 male-female pairs of *A. americanum.* Treatment cows were infested with 500 male-female pairs of ticks. The number of ticks attached to each cow and to each calf was counted 3 days later and the calves were reweighed after 7, 14, and 28 days.

In the analysis of data, each trial was treated as an independent event. ADG for each calf was corrected for autocorrelation among trials (Johnston 1984). To define tick damage, we fit a model of the general form (SAS 1985):

$$Y_i = \alpha + \beta_0 X_i,$$

where Y_i represent ADG for calf$_i$ and X_i is the density of attached female ticks on calf$_i$. β_0 estimates the effect of each female tick on ADG in a calf. α estimates ADG in the absence of tick infestation.

Fitted damage model coefficients (± SE) are α: 0.95079 (± 0.07191) and β_0: − 0.00176 (0.00048); R^2 = 0.955, *df* = 109. Our presumption of a functional relationship between ADG in calves and tick density is shown to be correct.

The damage model allows us to estimate lost productivity in a calf (1.76 g) caused by each feeding female tick. Most importantly, the damage model provides a basis for assessing the economic significance of *A. americanum* infestations on preweaner beef cattle.

3 Economic Impact of Chemically Mitigated *Amblyomma americanum* Populations on Preweaner Beef Cattle

That a demonstrable relationship exists between tick density and ADG in preweaner *B. taurus* cattle provides the biological rationale for the management of *A. americanum* on cattle. The economic rationale derives from a balance between the costs of tick control and the value of increased production, i.e., beef gain. Tick density at the intersection of these two functions is termed the economic injury level (Stern et al. 1959).

Four factors must be known to identify the economic injury level (Pedigo et al. 1986), vis-à-vis., market value, management costs, injury per tick, and host response to tick feeding. Injury per tick and host response to tick feeding are defined by the damage function.

Management costs comprise expenses for materials and labor. In developing the economic model, we considered three acaricides that can be applied topically to cattle: dioxathion, stirofos, and toxaphene + lindane. These acaricides reduce parasitic tick populations by 95%, 98%, and 94%, respectively, at 24 h post-treatment (Barnard and Jones 1981). A regimen of one acaricide treatment every 10 days for 100 days is used.

Labor costs for gathering and treating 25 cow-calf pairs ten times is estimated for three pasture sizes, 32, 65, and 97 ha, and three pasture conditions, good, average, or poor (Table 1). Good pastures have few vegetative or physio-

Table 1. Summary of costs (US$) of ten acaricide treatments applied to one cow-calf pair in a herd of 25 cow-calf pairs over a 100-day period (Barnard et al. 1986)

Pasture condition	Acaricide	Pasture size (ha)		
		32	65	97
Good	Dioxathion	6.06	6.38	6.38
	Stirofos	8.08	8.40	8.40
	Toxaphene + lindane	4.92	5.24	5.24
Average	Dioxathion	6.38	8.54	9.44
	Stirofos	8.40	10.56	11.45
	Toxaphene + lindane	5.24	7.40	8.29
Poor	Dioxathion	8.54	11.22	13.91
	Stirofos	10.56	13.24	15.93
	Toxaphene + lindane	7.40	10.08	12.77

graphic impediments to the gathering of livestock. Poor pastures are ≥ 30% brush per ha and have ≥ 1 physiographic impediment(s) per 5 ha. Average pasture conditions are between good and poor.

We used a product value of US$1.54/kg live-weight (the price received for beef calves in Oklahoma in December 1984). Growth in calves is considered for the 100-day period used for acaricide treatments.

To identify the economic injury level, four tick control scenarios are visualized. These scenarios are illustrated in Fig. 1 as: (A) cattle are tick free and acaricides are not used; (B) cattle are tick free but acaricides are used; (C) cattle are tick infested and acaricides are used; and (D) cattle are tick infested and acaricides are not used.

Scenario D manifests the tick damage function. The impact of tick damage on scenarios A, B, and C is dynamic and relates to fluctuating market prices for beef. This relationship is illustrated in Fig. 2 as the loss of beef returns from tick infestation that are associated with a variety of market conditions. The analysis of damage responses presented here is based on a US$1.54/kg live weight market value for beef calves.

Relative to the value of beef gain, scenario A (Fig. 2) is the optimal condition because it shows the value of added gain in the absence of ticks. The value of beef gain lost to the costs of tick control is the difference between scenarios A and B (shown as the shaded area). The difference between scenarios B and C is the value of beef gain lost as the result of < 100% acaricide efficacy. Scenario C is the actual condition. The area between scenarios C and D is the returns to the control measure above the costs of control, at different tick densities. The tick density at which the value of beef gain equals the costs of tick control (the intersection of scenarios C and D) is the economic injury level (Fig. 2). For this example (US$70/45 kg live weight and using stirofos for tick control), the economic injury level is 38 female ticks per calf.

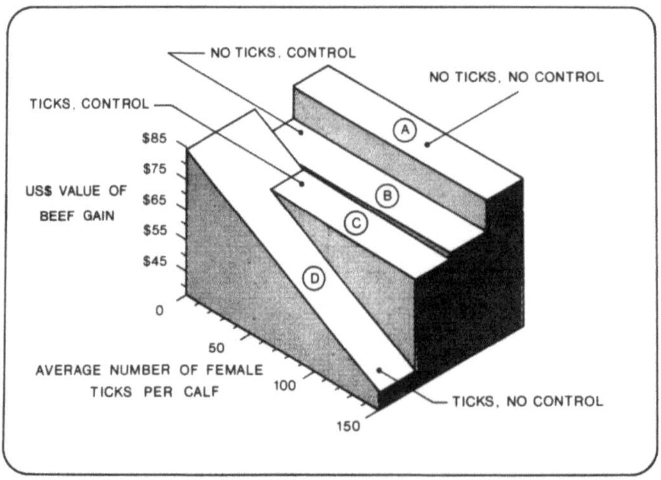

Fig. 1. Value added to *Bos taurus* beef cattle at different levels of beef production management. **A** Value of gain when calves are not tick infested and tick control is not applied. **B** Value of gain without ticks but with tick control. **C** Value of gain with ticks and with tick control. **D** Value of gain with ticks and without tick control

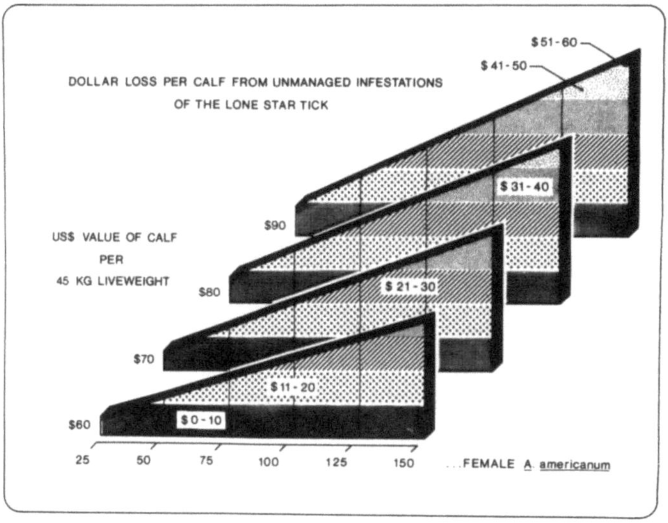

Fig. 2. Value loss per calf from unmanaged infestations of *Amblyomma americanum* at different calf values

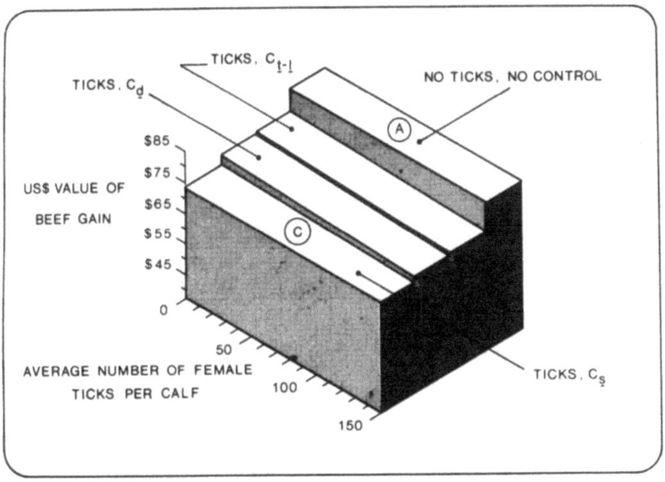

Fig. 3. Cost-effectiveness of acaricides used to manage infestations of *Amblyomma americanum* on beef calves. C_{t-l} Toxaphene + lindane; C_d dioxathion; C_s stirofos

Further analysis leads to a ranking of acaricides by cost and efficacy, relative to tick damage and the returns to beef gains. Using scenario C, the value of beef gain when cattle are tick infested and control measures are applied, we define the cost-efficiency ratio of dioxathion, stirofos and toxaphene + lindane relative to scenario A (Fig. 3). The most cost-efficient acaricide has the least distance (area) between itself and scenario A; here, toxaphene + lindane (C_{t-l}) is the most cost effective acaricide, followed by dioxathion (C_d) and then stirofos (C_s). Corresponding economic injury levels are 25.5, 29.8, and 38.0 female *A. americanum* per calf.

In sum, the use of acaricides for tick control on *B. taurus* cattle is economically feasible under certain conditions if daily populations of feeding female *A. americanum* on calves average 26–38 ticks. The main determinant of viability, according to our model, is labor costs. These costs are related directly to the condition and size of the pasture. With toxaphene + lindane, the cost of tick control for 100 days on an average 65 ha pasture (Table 1) is $7.40 per cow-calf unit. For dioxathion the cost is $8.54, and for stirofos the cost is $10.56. In each case, control costs, using a specific acaricide, must be lower than the resulting value of beef gain for the use of that acaricide against *A. americanum* to be economically feasible.

4 Economic Impact of Nonchemically Mitigated *Amblyomma americanum* Populations on Preweaner Beef Cattle

A number of entities affect calculation of the economic injury level (Pedigo et al. 1986). The primary model factors (market value, management costs, the damage function) are, themselves, dynamic. However, documented damage functions for arthropod pests of livestock are rare. The perception is that these pests do not readily evidence a quantitative relationship between injury (the deleterious effects of feeding) and damage (measurable loss of productivity) (Pedigo et al. 1986).

In fact, the principal constraint to defining economic injury levels for non chemical control of *A. americanum* is an inability to predict market value and management cost factors for each technology, relative to improved yield in beef cattle. This is true even for some chemical control methods. For example, returns from beef cattle to control measures such as acaricide impregnated slow release devices or systemic acaricides are unknown. Similarly, the economics of area treatment of pastures with acaricides is unknown.

This dilemma exists for vegetation management and animal management methods of tick control. Mechanical clearing of vegetation, burning, and the use of herbicides are labor intensive processes, but the expected returns are unknown. Meyer et al. (1982) analyzed dry matter production in pastures relative to mechanical clearing of tick habitat for control of *A. americanum*. Mechanical clearing is expensive, about US$350/ha, but the expense is mitigated by improved dry matter production at a low cost (US$0.14) per kilogram, and a concomitant reduction in the tick burden on cattle. The value of improved forage production and reduced tick burden as it relates to increased yield in beef cattle, however, has not been established.

Ridding a pasture of tick habitat with herbicides may improve the rate of dry matter production (Barnard 1985b), and at a lower cost per hectare (US$200) than for clearing. Mechanical flaming costs US$526/ha (Barnard 1985b) but free-living *A. americanum* populations are highly dispersed and only small areas of the pasture would require treatment. The impact of these control methods on populations of *A. americanum* parasitizing beef cattle has not been measured. Nevertheless, this impact must be quantified for returns from beef cattle to each control measure to be estimated. The long term practical and economic effects of vegetation management on forage crop production must also be understood.

The economic efficiency of host exclusion, pasture rotation, and herd management routines that are designed to reduce *A. americanum* populations on pastures, has not been analyzed and there are no field studies in which resistance to *A. americanum* in beef cattle has been shown to regulate tick population density. Efficacy data for host exclusion and herd manipulation methods of tick control are unavailable. And because the initial outlays of capital for labor and cattle handling equipment in these cases will be large, the risks and benefits to the beef producer must be carefully projected for 5–10 years, or longer, to facilitate user acceptance.

5 An Economics-Based Alternative Management Strategy for *Amblyomma americanum* on Beef Cattle

Managing populations of *A. americanum* on *Bos taurus* cattle with acaricides can provide the beef producer an economic advantage. With *B. indicus* cattle, this approach may be ineffectual because calf growth is not significantly compromised by tick infestation (Ervin et al. 1987).

B. indicus cattle present a possible alternative to acaricides for tick control. However, we do not know if the disparity in gain between tick-infested *B. indicus* and *B. taurus* cattle is sufficient to favor selection of a given breed type when data are analyzed within the whole farm context.

To study this question, we compared the economics of *B. indicus* and *B. taurus* calf production systems on pastures infested with *A. americanum*. The damage imposed by tick feeding on cattle is considered. A mathematical programming model is used to estimate returns to tick-infested pastures from cow-calf beef production systems with different breed types, and to identify levels at which equivalent returns are achieved for each breed type (Epplin and Ervin 1988).

Initial assumptions for the model are detailed in Table 2. Differences in pregnancy rate, calf survival, weaning rate, and weaning weight among breed types cause variation in the kilograms of calf weaned. Differences in feed efficiency and market price influence the levels of net returns. Of necessity our base estimates of production coefficients for these factors have been acquired from a variety of sources (Table 2).

The assumption of 7.3% greater weaning weights for *B. indicus* cattle, and 11.2% reduction in gain over the 120-day-long adult tick season for *B. taurus* cattle, results in 13.6% more pounds of calf per cow for the *B. indicus* cattle. We assumed 4.7% more feed is required per *B. indicus* cow unit, and we assumed that *B. indicus* steers and heifers are discounted at market by 2.35 and 0.61%, respectively (Lambert et al. 1983).

Returns were computed by breed type and we considered both spring and fall calving. Forage production in the spring is amenable to high stocking

Table 2. Assumptions regarding differences in weaning weight, market price, feed requirements, and tick damage between *Bos indicus* and *B. taurus* (Epplin and Ervin 1988)

Initial assumption
Weaning weight
7.3% greater weaning weight for *B. indicus* (Cundiff et al. 1975; Crockett et al. 1978a, b; USDA 1978; Bolton et al. 1987)
Market price
2.35% price discount for *B. indicus* steers 0.61% price discount for *B. indicus* heifers (USDA 1978; Lambert et al. 1983)
Feed requirements
4.7% more metabolizable energy and digestible protein for *B. indicus* cow unit (Frisch and Vercoe 1977; Crockett et al. 1978a; USDA 1978; NRC 1984; Frahm and Marshall 1986)
Tick damage
11.2% reduction in calf gain over tick season for *B. taurus* steer and heifer calves (Barnard et al. 1986; Ervin et al. 1987)

densities of cattle and results in the greatest number of calves produced per unit of land. We found spring calving to be more economical than fall calving and considered only spring calving in subsequent analyses.

Utilizing base assumptions, the *B. indicus* system with spring calving generates a return to the fixed resources of \$25.48/ha of tick infested pasture. When the *B. indicus* system with spring calving is eliminated from the model, the *B. taurus* system with spring calving generates a return to the fixed resources of \$21.42/ha.

For price parameterization, *B. indicus* steer and heifer prices were decreased simultaneously to locate parameter values at which returns would be equivalent between breed systems. This was done because the rate at which both prices should be proportionately decreased is unknown. Two different sets of assumptions are therefore used to derive the prices at which producers earn equivalent returns from each breed type on tick infested pastures. In the first analysis, the ratio of *B. indicus* heifer price to *B. indicus* steer price (Table 3: [\$1.416 ÷ \$1.698]) is held constant at 0.83. In this case, *B. indicus* cattle are preferred when the price exceeds \$1.562 for *B. indicus* steers and \$1.303 for heifers. The *B. indicus* system would be more economical when the price discount, relative to *B. taurus* cattle, was less than 10.1% for steers and 8.5% for heifers.

In the second analysis, the initial ratio at which *B. indicus* heifers and steers are discounted, relative to *B. taurus* cattle (Table 3: ([1 − {\$1.416 ÷ \$1.425}] ÷ (1 − [\$1.698 ÷ \$1.738])), is held constant at 0.26. Here, the *B. indicus* system generates more returns than the *B. taurus* system if the price discounts are less than 12.7 and 3.9% for *B. indicus* steers and heifers, respectively.

To identify the weaning weight difference at which both *B. indicus* and *B. taurus* breed types return equivalent amounts per unit of land, we simultaneously decreased the average pounds (weight) of steer and heifer calf produced for sale, per cow unit, for the *B. indicus* cattle (Table 4). If *B. indicus* cattle produce 4.6% more pounds of calf weaned per cow unit than the *B. taurus* cattle, the two breed types would generate equivalent returns.

Table 3. Break-even prices (US$ per 0.45 kg) for *Bos indicus* cattle on pastures infested with *Amblyomma americanum*

Bos indicus		*Bos taurus*	
Steers	Heifers	Steers	Heifers
Base prices			
1.698 (0.977)[a]	1.416 (0.994)[a]	1.738	1.425
Break-even prices[b]			
1.562 (0.899)[a]	1.303 (0.915)[a]	1.738	1.425
Break-even prices[c]			
1.517 (0.873)[a]	1.369 (0.961)[a]	1.738	1.425

[a] *Bos indicus/Bos taurus* price ratio.
[b] Assuming a *Bos indicus* heifer/steer price ratio of 0.83 ($1.416 ÷ $1.698).
[c] Assuming a *Bos indicus* heifer/steer price discount ratio of 0.26 (0.61% ÷ 2.35%: Table 2).

Table 4. Base weaning weights and calculated break-even weights for steers among breed types

Bos indicus	*Bos taurus*	Ratio[a]
Base weaning weights (kilograms)		
210.20	195.90	1.073
Base weaning weights (kilograms) adjusted for ticks		
210.20	185.04	1.136
Weaning weights (kilograms) resulting in equivalent returns among breed types		
193.74	185.04	1.047

[a] *Bos indicus* ÷ *Bos taurus.*

Each feed requirement coefficient for the *B. indicus* cattle was simultaneously increased in fixed proportions. The *B. indicus* cattle system is most economical until the feed requirements per cow unit are 16.5% greater than for the *B. taurus* system (the base model includes only 4.7% more feed required for the *B. indicus* cattle).

Results of sensitivity analyses are summarized in Table 5. When discounts for *B. indicus* steer and heifer prices are tripled (from 2.35 and 0.61% to 7.05 and 1.83%, respectively), the corresponding *B. indicus* to *B. taurus* ratios are 0.929 and 0.982, and the *B. indicus* cattle generate 10.6% more returns per unit of tick infested pasture. Tripling the price discount reduces the break-even feed requirement ratio from 1.165 to 1.113 and increases the break-even weight sold ratio from 1.046 to 1.084. Both estimates are different from the base assumption ratios of 1.047 for feed requirement and 1.136 for weight sold (Table 5).

To summarize, we found that the *B. indicus* production system generated the highest returns to the resource base and 19% more returns per hectare to tick infested pasture than the *B. taurus* production system. Sensitivity analyses

Table 5. Ratios (*Bos indicus*/*Bos taurus*) for market price, feed requirements, and weight sold resulting in equivalent returns to both breed types for *Amblyomma americanum* infested pastures

Steer price	Heifer price	Weight sold	Feed consumption	Returns
Base assumptions[a]				
0.977	0.994	1.136	1.047	1.190
		Alternative break-even ratios		
0.899[a]	0.915	1.136	1.047	1.000
0.873[b]	0.961	1.136	1.047	1.000
0.977	0.994	1.136	1.165	1.000
0.977	0.994	1.046	1.047	1.000
Price discount for *Bos indicus* cattle doubled from base				
0.953	0.988	1.136	1.047	1.174
		Alternative break-even ratios		
0.953	0.988	1.136	1.139	1.000
0.953	0.988	1.065	1.047	1.000
Price discount for *Bos indicus* cattle tripled from base				
0.929	0.982	1.136	1.113	1.000
		Alternative break-even ratios		
0.929	0.982	1.136	1.113	1.000
0.929	0.982	1.084	1.047	1.000

[a] Assuming the *Bos indicus* heifer/steer price ratio = 0.83.
[b] Assuming the *Bos indicus* heifer/steer price discount ratio = 0.26.

indicate that the use of *B. indicus* cattle on tick infested pasture is economically sound over the ranges considered for each parameter.

Practical application of the findings of this study cannot be made without some qualification. Constraints exist to the acceptance of *B. indicus* cattle by beef producers. Purported behavioral and reproductive properties and carcass quality are the principal issues. In fact, *B. indicus* cattle reach sexual maturity later than *B. taurus* cattle; however, *B. indicus* cattle have greater reproductive longevity. On a lifetime basis *B. indicus* cattle may be more fertile than *B. taurus* cattle (Cartwright 1980).

With respect to carcass quality, *B. indicus* cattle do not grade choice at typical slaughter weights. However, if fed to weights to achieve choice grade they are discounted because of heavy carcass weight. Despite this handicap, modelling results suggest that *B. indicus* cattle can generate greater returns to tick infested pastures than *B. taurus* cattle.

This study does have limitations. First, it is not a comprehensive analysis of all potential uses of tick infested pasture, but a comparison of alternative cow-calf production activities. Estimates of liveweight gain difference among breed types in other kinds of beef production activities such as "backgrounding" (Neumann 1977) are not available.

Second, alternative tick control measures are not considered. Chemicals can be used. Barnard et al. (1986) reported the labor and materials costs of a control program for *A. americanum* on beef cattle to range from US$7.40 to 12.77. At the prices budgeted in this analysis, the value of weight loss per *B. taurus* cow unit is US$10.88. The use of acaricides thus exhibits marginal economic viability and we did not consider production alternatives that included chemical control. As noted before, management costs for non-chemical tick control methods relevant to parasitic *A. americanum* populations do not exist.

A third limitation of the study is that potential differences in investment requirements and replacement rates among breed types of cattle are not considered. The impact of a *B. indicus* production system, relative to a *B. taurus* system, on variations in investment requirements for cows, bulls, and livestock equipment is equivocal. Estimates of the effect of *B. indicus* cattle on fencing and corral costs are not available. And there are no estimates of relative differences in labor requirements based on breed types of cattle.

Finally, we have no data for tick damage in cattle that have varying levels of *B. indicus* breeding. If future research indicates a negative relationship between tick damage and increasing levels of *B. indicus* breeding, the incorporation of *B. indicus* genotypes into beef cattle breeding programs in the US has the potential to raise the economic injury level for *A. americanum*. And while maintaining consistent levels of *B. indicus* breeding may be an untenable goal for most farms in the US, composite breeds (Braford, Brangus, etc.) may be appropriate substitutes in these situations (Gregory and Cundiff 1980).

References

Barnard DR (1981) *Amblyomma americanum*: a comparison of populations of ticks free living on pastures and parasitic on cattle. Ann Entomol Soc Am 74:507–511

Barnard DR (1985a) Injury thresholds and production loss functions for the lone star tick, *Amblyomma americanum* (Acari: Ixodidae) on pastured preweaner beef cattle, *Bos taurus*. J Econ Entomol 78:852–855

Barnard DR (1985b) Density perturbation in populations of *Amblyomma americanum* (Acari: Ixodidae) in beef cattle forage areas in response to two regimens of vegetation modification. J Econ Entomol 79:122–129

Barnard DR (1988) Habitat use by cattle affects contact with lone star ticks. J Econ Entomol 82:854–859

Barnard DR, Jones BG (1981) Field efficacy of acaricides for control of the lone star tick on cattle in southeastern Oklahoma. J Econ Entomol 74:558–560

Barnard DR, Ervin RT, Epplin FM (1986) Production system-based model for defining economic thresholds in preweaner beef cattle, *Bos taurus*, infested with the lone star tick, *Amblyomma americanum* (Acari: Ixodidae). J Econ Entomol 79:141–143

Barnard DR, Mount GA, Koch HG, Haile DG, Garris GI (1989) Management of the lone star tick in recreation areas. United States Department of Agriculture. Agricultural Research Service. Agricultural Handbook 682. U.S. Government Printing Office, Washington, DC

Bloemer SR, Snoddy EL, Cooney JC, Fairbanks K (1986) Influence of deer exclusion on populations of lone star ticks and American dog ticks. J Econ Entomol 79:679–683

Bolton RC, Frahm RR, Castree JW, Coleman SW (1987) Genotype × environment interactions involving proportion of Brahman breeding and season of birth. I. Calf growth to weaning. J Anim Sci 65:42–47

Cartwright TC (1980) Prognosis of Zebu cattle: research and application. J Anim Sci 50:1221–1226

Clymer BC, Howell DC, Hair JA (1970) Environmental alteration in recreational areas by mechanical and chemical treatment as a means of lone star tick control. J Econ Entomol 63:504–509
Cooley RA, Kohls GM (1944) The genus *Amblyomma* in the United States. J Parasitol 30:77–118
Crockett JR, Koger M, Franke DE (1978a) Rotational crossbreeding of beef cattle: reproduction by generation. J Anim Sci 46:1163–1169
Crockett JR, Koger M, Franke DE (1978b) Rotational crossbreeding of beef cattle: preweaning traits by generation. J Anim Sci 46:1170–1171
Cundiff LV, Gregory KE, Koch RM (1975) Characterization of breeds representing diverse biological types: reproduction and maternal performance of F_1 cows. ARS Techn Publ 42. U.S. Government Printing Office; Washington, DC
Drummond RO (1967) Seasonal activity of ticks (Acarina: Metastigmata) on cattle in southwestern Texas. Ann Entomol Soc Am 60:439–447
Drummond RO, Lambert G, Smalley HE Jr, Terril CE (1981) Estimated losses of livestock to pests. In: Pimentel D (ed) CRC Handbook of pest management in agriculture, vol 1. CRC, Boca Raton
Drummond RO, George JE, Kunz SE (1988) Control of arthropod pests. A review of technology. CRC, Boca Raton
Epplin FM, Ervin RT (1988) Evaluating the use of an alternate breed type as a pest management strategy. North Cent J Agric Econ 10:225–232
Ervin RT, Epplin FM, Barnard DR (1985) Estimates of costs inflicted by the lone star tick on calves in eastern Oklahoma. Curr Farm Econ 58:9–14
Ervin RT, Epplin FM, Byford RL, Hair JA (1987) Estimation and economic implications of lone star tick (Acari: Ixodidae) infestation on weight gain of cattle, *Bos taurus* and *Bos taurus* × *Bos indicus*. J Econ Entomol 80:443–445
Frahm RR, Marshall DM (1986) Comparisons among two breed cow groups. I. Cow productivity and calf performance to weaning. J Anim Sci 61:844–855
Frisch JE, Vercoe JE (1977) Food intake, eating rate, weight gains, metabolic rate, and efficiency of feed utilization in *Bos taurus* and *Bos indicus* crossbred cattle. Anim Prod 25:343–358
George JE, Osburn RL, Wikel SK (1985) Acquisition and expression of resistance by *Bos indicus* and *Bos indicus* × *Bos taurus* calves to *Amblyomma americanum*. J Parasitol 7:174–182
Gregory KE, Cundiff LV (1980) Crossbreeding in beef cattle: evaluation of systems. J Anim Sci 51:1224–1242
Hair JA, Howell DE (1971) Lone star ticks: their biology and control in Ozark recreation areas. Okla Agric Exp Stn Bull 679
Hoch AL, Barker RW, Hair JA (1971) Further observations on the control of lone star ticks (Acarina: Ixodidae) through integrated control procedures. J Med Entomol 8:731–734
Hughes JG (1976) Short-term variation in animal liveweight and reduction of its effect on weighing. Anim Breed Abstr 44:111–118
Johnston J (1984) Econometric methods. McGraw-Hill, New York
Lambert C, Corah L, Grunewald O (1983) Factors affecting processing of calves and yearlings in Kansas. Kansas State University Report MF-667 Manhattan
Meyer JA, Lancaster JL, Simco JS (1982) Comparison of habitat modification, animal control, and standard spraying for control of the lone star tick. J Econ Entomol 75:524–529
Neumann AL (1977) Beef cattle. Wiley, New York
NRC (1984) National Research Council. Nutrient requirements of beef cattle. National Academy Press, Washington, DC
Pedigo LP, Hutchins SH, Highley LG (1986) Economic injury levels in theory and practice. Annu Rev Entomol 31:341–368
Rechav Y, Zeederberg ME (1986) Tick populations on two breeds of cattle under field conditions, with a note on blood components related to host resistance. In: Sauer JR, Hair JA (eds) Morphology, physiology, and behavioral biology of ticks. Ellis Horwood, Chichester, pp 445–456
SAS Institute Inc. (1985) Statistical analysis system user's guide: statistics. SAS Institute, Cary, NC
Smith RF, van den Bosch R (1967) Integrated control. In: Kilgore WW, Doutt EL (eds) Pest control. Biological, physical, and selected chemical methods. Academic Press, New York, pp 295–324
Stern VM, Smith RF, van den Bosch R, Hagen KS (1959) The integration of chemical and biological control of the spotted alfalfa aphid. Part 1. The integrated control concept. Hilgardia 29:81–101

Stone JD, Pedigo LP (1972) Development and economic injury levels of the green cloverworm on soybean in Iowa. J Econ Entomol 65:197–201
Strother GR, Burns EC, Smart LI (1974) Resistance of purebred Brahman, Hereford, and Brahman × Hereford crossbred cattle to the lone star tick, *Amblyomma americanum* (Acarina: Ixodidae) J Med Entomol 11:559–563
Sutherst RW, Utech KBW (1981) Controlling livestock parasites with host resistance. In: Pimentel D (ed) CRC Handbook of pest management in agriculture. CRC, Boca Raton, pp 385–407
Sutherst RW, Maywald GF, Bourne AS, Sutherland ID, Stegeman DA (1988) Ecology of the cattle tick (*Boophilus microplus*) in subtropical Australia. II. Resistance of different breeds of cattle. Aust J Agric Res 39:299–308
USDA (1978) U.S. Department of Agriculture germ plasm evaluation program. Prog Rep No 6, ARM-NC-2. U.S. Government Printing Office, Washington, DC
USDA (1988) Agricultural statistics. United States Department of Agriculture. U.S. Government Printing Office, Washington, DC
Utech KBW, Wharton RH, Kerr JD (1978) Resistance to *Boophilus microplus* (Canestrini) in different breeds of cattle. Aust J Agric Res 29:885–895
Wilkinson PR (1957) The spelling of pastures in cattle tick control. Aust J Agric Res 8:414–423
Wilkinson PR (1977) Effect of herbicidal killing of shrubs on abundance of adult *Dermacentor andersoni* (Acarina: Ixodidae) in British Columbia. J Med Entomol 13:713–718

Biological Processes in the Epidemiology of Heartwater

R.A.I. Norval[1], H.R. Andrew[2], C.E. Yunker[2] and M.J. Burridge[1]

Summary

Heartwater, a rickettsial disease of ruminants caused by *Cowdria ruminantium*, is transmitted by various tick species of the genus *Amblyomma*. It occurs through most of sub-Saharan Africa, where it is of considerable economic importance, and has spread to a number of islands in the Indian and Atlantic Oceans and the Caribbean and now poses a threat to the American mainland.

Until recently our understanding of the epidemiology of the disease was poor. However, this situation is changing with developments such as the *in vitro* culture of *C. ruminantium* and increased research on heartwater and its vectors in a number of countries. The purpose of this chapter is to review our knowledge of various biological processes in the epidemiology of heartwater, including host/tick interactions, the persistence of infection in ruminants and infection in vectors. Background information on the ecology and distribution of the vectors and on the disease itself is also given. The control of heartwater is discussed in the light of new information.

1 Introduction

Heartwater is a rickettsial disease of ruminants, caused by *Cowdria ruminantium*. Its original distribution appears to have been sub-Saharan Africa but in the past centuries it has spread to various other localities (islands) in the tropics. Its vectors are a number of tick species of the genus *Amblyomma*.

In terms of tick-borne diseases of cattle in Africa, heartwater is surpassed in importance only by East Coast fever (ECF), caused by the *Theileria parva* group of organisms (Uilenberg 1981). In the areas in which ECF does not occur (western and southern Africa) heartwater is regarded as the most important tick-borne disease of cattle (Norval 1989). Unlike the other commonly occuring tick-borne diseases of cattle, it can also cause losses in other domestic and some wild ruminant species.

Heartwater is seldom a problem in indigenous livestock in endemic areas because young animals acquire immunity to the disease through natural exposure (Neitz and Alexander 1941; Thomas and Mansvelt 1957; Uilenberg 1983a; Norval 1989). The circumstances in which heartwater causes losses in domestic livestock have been reviewed extensively (Camus and Barre 1982;

[1] Center for Tropical Animal Health, College of Veterinary Medicine, University of Florida, Gainesville, Florida 32610, USA
[2] University of Florida/USAID/Zimbabwe Heartwater Research Project, Veterinary Research Laboratory, P.O. Box 8101, Causeway, Zimbabwe

Uilenberg 1983a; Provost and Bezuidenhout 1987; Bezuidenhout and Bigalke 1987). These include situations where:

1) stock, both indigenous and exotic, are moved from heartwater-free to endemic areas;
2) highly susceptible stock (e.g. Angora goats) are raised in endemic areas;
3) conditions are ecologically marginal for vector survival and not all stock are immunized by natural infection when young;
4) intensive tick control is practised and the disease is suppressed but not eradicated; and
5) vectors and the disease spread to new areas.

Heartwater is becoming increasingly important because of changing agricultural practices in Africa, including the more frequent use of imported exotic breeds of livestock to improve productivity, extension of intensive livestock farming into areas that are ecologically marginal for vector survival and increasing movement of livestock between heartwater-free and endemic areas. Another cause of concern is the potential for the spread of the disease in Africa (Norval 1983; Andrew 1990) and in the western hemisphere (Barre et al. 1987).

The history of heartwater, including the discovery of its transmission by *Amblyomma* ticks, the identification of the causal organism, the development of diagnostic techniques, the study of acquired immunity to the disease and immunization using live organisms, has been comprehensively reviewed by Camus and Barre (1982). After the initial important discoveries, mostly in South Africa during the first half of the century, few advances were made in our understanding of the epidemiology and control of heartwater until the 1980s. This was due in part to the relatively small amount of research that was conducted on heartwater in the 1960s and 1970s and, in part, to problems associated with the propagation of *C. ruminantium* in vitro and in vivo in laboratory hosts, as sources of antigen for diagnostic tests and other purposes.

The progress made in heartwater research subsequent to 1980 occurred for several reasons. There has been renewed interest in the disease as its importance has been recognised in an increasing number of countries and situations; as a consequence more research has been carried out in countries other than South Africa. The growth of *C. ruminantium* in mice, initially reported by Du Plessis and Kumm (1971), was repeated by Mackenzie and Van Rooyen (1981). Several stocks of *C. ruminantium* have been grown in vitro in bovine endothelial cell cultures (Bezuidenhout et al. 1985; Bezuidenhout 1987a; Yunker et al. 1988) and caprine neutrophil cultures (Jongegan et al. 1986; Logan et al. 1987). Technological advances, including recombinant DNA techniques, have made feasible the development of new tools such as diagnostic DNA probes (Ambrosio et al. 1987). Research on *Amblyomma* pheromones has yielded a method of sampling the unfed nymphal and adult stages of the ticks (Norval et al. 1989a) and has produced new insights into vector ecology and the epidemiology of heartwater (Norval et al. 1989b).

The aims of this chapter are to review aspects of the epidemiology of heartwater and to discuss the significance and implications of recent findings. Topics covered are the disease and its expression in different hosts; the carrier state and the likely roles of domestic and wild vertebrates as reservoirs; the distribution, ecology and behaviour of the vectors; infection in vectors and the importance of unfed ticks and feeding males as reservoirs. The control of the disease is discussed in the light of the new knowledge.

2 Heartwater Disease

The distribution of heartwater corresponds closely with that of its African vectors and includes sub-Saharan Africa and at least some of the islands to which *A. variegatum* has spread (Provost and Bezuidenhout 1987). The disease does not occur on the American mainland but there is a danger that it could be introduced from adjacent Caribbean islands and become established (Barre et al. 1987).

Little has been published on the seasonal occurrence of heartwater but it is generally recognized that cases can occur at any time that vectors are present (Gruss 1987). In Guadeloupe, where *A. variegatum* shows no seasonality, the disease occurs throughout the year (Camus and Barre 1987).

Heartwater can manifest itself clinically in varying forms and severity depending on the age, breed, species and immune status of ruminants affected. Clinical signs in susceptible animals include fever, inappetence, incoordination, respiratory distress, nervous symptoms and death (Van de Pypekamp and Prozesky 1987). Macroscopic pathological lesions are related to increased capillary permeability, which leads to the excess effusion of fluid into tissues and the body cavities (Clark 1962; Prozesky and Du Plessis 1984). Thus, at autopsy varying combinations and degrees of hydrothorax, hydropericardium (hence the name of the disease), ascites, and brain, lung, mediastinal, abomasal and lymphnode oedema may be observed. Nephrosis, splenomegaly and enterorrhagia may also occur.

Calves under the age of three weeks, irrespective of the immune status of the dam, possess an innate resistance to heartwater (Neitz and Alexander 1941; Alexander et al. 1946). Lambs of all breeds younger than 8 days and kids of less than 6 weeks are also usually resistant to the disease (Neitz and Alexander 1941; Thomas and Mansvelt 1957).

Breed resistance to heartwater in cattle, sheep and goats has been reviewed by Matheron et al. (1987). In general, local or indigenous breeds developed in endemic areas are more resistant than exotic breeds. There is some evidence that resistance in a population is due to selection through exposure to the disease.

The occurrence of heartwater in wild ruminants has been reviewed by Oberem and Bezuidenhout (1987a). A number of African and non-African species have been found to be susceptible to the disease and show symptoms similar to those in domestic ruminants. In several other species, including the African buffalo, no symptoms have been reported.

3 Carrier State

Until very recently a carrier state for heartwater was not thought to exist in domestic ruminants. It was generally accepted that animals were infective during the febrile reaction and for a short period thereafter (reviews by Camus and Barre 1982; Uilenberg 1983a; Barre and Camus 1987).

Wildlife was considered, on the basis of largely circumstantial evidence, to be involved in the epidemiology of heartwater. A number of ruminant and other vertebrate species, including a reptile, a bird, a rodent and a primate, had been shown to become infected with *C. ruminantium* (reviews by Camus and Barre 1982; Uilenberg 1983a; Oberem and Bezuidenhout 1987a) and the blood of two ruminant species, blesbuck (*Damaliscus dorcas*) and black wildebeest (*Connochaetes gnu*), had been found to be infective (Neitz 1935). In the cases of the reptile (tortoise, *Geochelone pardalis*) and bird (helmeted guinea fowl, *Numida meleagris*) it was also shown that infection could be acquired and transmitted by *Amblyomma* ticks (Oberem and Bezuidenhout 1987a). However, it was not known if any wild hosts remained long-term, asymptomatic carriers of *C. ruminantium*. Also unknown was the epidemiological importance, if any, of those species that were refractory to infection (Camus and Barre 1982; Oberem and Bezuidenhout 1987a).

It has now been shown that domestic and wild ruminants can remain long-term, asymptomatic carriers of *C. ruminantium* (Andrew and Norval 1989b). Sheep were still infective for nymphs of *A. hebraeum* 223 days after inoculation; cattle, 246 days and African buffalo, 161 days. Previous studies, indicating the absence of the carrier state, involved attempts at transmission using nymphs fed as larvae on recovered animals or blood from recovered animals. Nymphs appear to be effective in picking up low levels of infection because, as shown by Kocan et al. (1987), *Cowdria* organisms multiply in their gut epithelial cells. Nymphs therefore have a concentrating or amplifying effect, which may not occur, or not to the same extent, in larvae. Larvae also ingest much smaller volumes of blood than nymphs and so are exposed to lower numbers of organisms. It appears that tick transmission from larva to nymph and needle transmission by inoculation of blood occur only when rickettsaemias are high, during and immediately after the febrile reaction.

The fact that domestic ruminants and buffalo can serve as reservoirs of infection is of considerable importance in the epidemiology of heartwater. It means that *Amblyomma* ticks feeding on asymptomatic hosts can become infected and that the numbers of vectors required for the perpetuation of the disease are almost certainly lower than were previously believed. It also means that wild hosts are potentially important reservoirs of infection; if buffalo, which were believed to be refractory hosts, can become carriers it seems likely that the same will be true for a variety of other susceptible and apparently refractory ruminant species. The importance of non-ruminants requires further study.

4 Vector Species

Information on the transmission of heartwater by ticks has been reviewed by Camus and Barre (1982), Uilenberg (1983a) and Bezuidenhout (1987b). In Africa, the known vectors are *Amblyomma astrion*, *A. cohaerens*, *A. gemma*, *A. hebraeum*, *A. lepidum*, *A. marmoreum*, *A. pomposum*, *A. sparsum*, *A. tholloni* and *A. variegatum*. Of these, the only one established beyond the offshore islands of Africa is *A. variegatum*. Two species from the western hemisphere, *Amblyomma cajennense* and *A. maculatum*, have been found to transmit the disease under experimental conditions (Uilenberg 1982, 1983b).

4.1 Distribution of Vectors

Recently, maps of the distributions of the *Amblyomma* vectors of heartwater have been compiled by Walker and Olwage (1987). In Africa, the most widely distributed species is *A. variegatum*, which occurs throughout west Africa from the Sahara southwards, in areas of north Africa including southern Sudan, most of Ethiopia and northwestern Somalia, in almost all of central and east Africa, and in southern Africa as far south as the Zambezi River system and central Mozambique. In highland areas to the south of the Congo basin, *A. variegatum* gives way to *A. pomposum* and in southern Africa it is replaced by *A. hebraeum*. There is evidence that *A. variegatum* and *A hebraeum* are mutually exclusive due to interspecific competition (Rechav et al. 1982; Norval 1983). *Amblyomma cohaerens*, *A. gemma* and *A. lepidum* occur fairly widely in east and northeastern Africa. *Amblyomma astrion* occurs in the Congo basin, in coastal areas of Angola and on the west African islands of Sao Tome and Principe. *Amblyomma sparsum* and *A. tholloni* have been collected from scattered localities over the greater part of sub-Saharan Africa, and *A. marmoreum* occurs at scattered localities in southern Africa. The only parts of sub-Saharan Africa from which *Amblyomma* ticks are absent are deserts and cold highlands areas.

Outside the African continent *A. variegatum* is established in Yemen, on the Indian Ocean islands of Madagascar, Grande Comore, Reunion and Mauritius, on the Atlantic Ocean islands of Cape Verde and on the Caribbean islands of Guadeloupe, Marie Galante, Martinique, Antigua, St. Lucia, Nevis, St. Kitts, St. Maarten, La Desirade, Monserrat and Dominica. It has also been recorded from or temporarily established on several other islands in the Antillean chain (Anguilla, Barbados, Puerto Rico, the US Virgin Islands and Viegues).

Amblyomma maculatum occurs primarily in coastal areas around the Gulf of Mexico. *Amblyomma cajennense* is more widely distributed, occurring through Central America, Cuba and much of South America as far south as Uruguay.

4.2 Ecology of Vectors

Information on the ecology of the African *Amblyomma* vectors of heartwater has been reviewed by Petney et al. (1987). All have three-host life cycles, mostly characterized by long development periods. In general, the immature stages have wide host ranges and the adults are fairly host specific. Petney et al. (1987) allocated the species to three groups on the basis of their host associations and hence importance as vectors of heartwater. The first group, comprising the main vectors, includes *A. hebraeum* and *A. variegatum*. Both these species are widely distributed and can feed exclusively on domestic animals, using cattle as the major hosts of the adults. The second group, comprising vectors of secondary importance, includes *A. astrion*, *A. cohaerens*, *A. gemma*, *A. lepidum* and *A. pomposum*. These species have either fairly limited distributions or occur only in small numbers on domestic stock. The third group, comprising accidental vectors, includes *A. marmoreum*, *A. sparsum* and *A. tholloni*. The adults of these have specific wild hosts, tortoises, tortoises/rhinoceroses (*Diceros bicornis*)/buffalo (*Syncerus caffer*) and elephants (*Loxodonta africana*) respectively, and seldom or never parasitize domestic stock. Although the nymphs of *A. sparsum* and *A. tholloni* can transmit heartwater, these species are seldom in contact with domestic animals because of the host requirements of their adults. *Amblyomma marmoreum*, which is common in farming areas in some parts of South Africa, may be of greater importance as a vector. Its immature stages occur frequently on domestic ruminants (Norval 1975; Horak and Knight 1986) and it has been shown that the tortoise *G. pardalis*, on which the immature stages of *A. marmoreum* and the nymphs of *A. hebraeum* frequently feed (Norval 1975; Walker and Schultz 1984), can be a source of infection for ticks (Oberem and Bezuidenhout 1987a).

With further investigation it may be found that *A. hebraeum* and *A. variegatum* are not the only major vectors of heartwater. In western Ethiopia, for example, "*A. cohaerens* is the most prevalent and abundant tick on cattle" (Pegram et al. 1981). When the livestock densities of the highlands of Ethiopia are compared with those of southern Africa it is apparent that *A. cohaerens* may transmit *C. ruminantium* to as many or more animals than *A. hebraeum*.

The abundance of *Amblyomma* ticks can vary considerably, with place and time. A number of factors affect abundance. One of the most important is the presence of suitable hosts for the adult and immature stages. The species that become most abundant in agricultural situations tend to be those that are able to feed on domestic animals in all stages (Norval 1979). The abundance of species that require specific hosts at any stage of their life cycle are obviously limited by the abundance of these hosts. Other factors of importance include acaricide usage, climate and vegetation. The marked effect that acaricide treatment of domestic animals can have on *Amblyomma* populations has recently been demonstrated in South Africa by Horak and Knight (1986) and Petney and Horak (1987) and in the Caribbean by Garris (1987). Cold climates are thought to adversely affect the survival of *A. hebraeum* by prolonging the development periods (Norval 1977b) and so altering patterns of seasonal occurrence (Norval 1983). The effect on other species is probably similar. Vegetation is important for the survival of the ticks as it provides shade and hence the microclimatic conditions necessary for the survival of the free-living stages (Norval 1977a). The African *Amblyomma* vectors of heartwater, including dry-adapted species such as *A. gemma* and *A. lepidum*, all require tree and grass cover for their survival (Petney et al. 1987).

Among the African *Ambylomma* species, detailed information on seasonal occurrence is available only for *A. hebraeum* and *A. variegatum* (Petney et al. 1987). The main herbivore parasites, however, appear to be similar to these species in that adults are most abundant during periods of rain, independent of the timing of the rains. The regulation of seasonal occurrence is important to ensure that the desiccation-sensitive eggs are laid when microclimatic conditions are most favourable for their survival (Norval 1977b). This regulation in *A. hebraeum* and *A. variegatum* appears to be achieved by two independent mechanisms; regulation of adult activity to coincide with the rains and diapause in engorged females which delays the start of oviposition until the onset of rains (Yeoman 1968; Norval 1977a; Rechav 1982; Pegram et al. 1988). The pattern of seasonal occurrence appears to be more clearly defined in *A. variegatum* than in *A. hebraeum*. This is illustrated by the fact that, in Zambia, adults of *A. variegatum* occur mainly between October and February, larvae only from March to May and nymphs from May to September (Pegram et al. 1986). In southern Zimbabwe all stages of *A. hebraeum* are present throughout the year and patterns of seasonal occurrence are not clearly defined (Norval et al. 1991).

Amblyomma hebraeum and *A. variegatum* normally pass through one generation each year. However, it has been suggested that there may be more than one generation of *A. variegatum* when there is more than one rainy season (Hoogstraal 1956). In contrast, there is evidence that *A. hebraeum* may sometimes require more than one year to pass through a generation in the extreme southern limits of its distribution (Norval 1977b; Rechav 1982).

The unfed nymphs and adults of *Amblyomma* ticks survive for extremely long periods under conditions of naturally fluctuating temperature and relative humidity. Lewis (1939) cited by Hoogstraal (1956) recorded the survival of *A. variegatum* for 732 days (Hoogstraal presumed that he was referring to unfed adults). In Zambia, Pegram and Banda (1989) found that, when confined to gauze columns, unfed nymphs of *A. variegatum* survived 12–24 weeks and unfed adults 25–33 weeks. In South Africa, Norval (1977b) recorded 50% survival times of + 56–63 and 32–46 weeks for unfed nymphs and adults of *A. hebraeum* respectively, in a Stevenson screen. One can assume that, unconfined, the ticks would have survived for longer, in the protected microhabitats in which they normally occur, on or below the soil surface (Norval et al. 1987, 1988a).

4.3 Behaviour of Vectors

The larvae of *A. hebraeum* and *A. variegatum*, like those of many other ixodid tick species, ascend vegetation to await passing hosts (Rechav 1979; Pegram and Banda 1989). The nymphs and adults, on the other hand, shelter beneath the debris on the soil surface and become active only in response to specific host stimuli (Norval et al. 1987, 1988a, 1989a; Pegram and Banda 1989).

Until recently little was known about host selection or host location by the unfed nymphs and adults. It is now known that the ticks are activated by carbon dioxide (Norval et al. 1987, 1988a) and that the directional response necessary for host location is provided by the aggregation-attachment pheromone (AAP), which is produced by attached males (Hess and de Castro 1986; Norval et al. 1989a, b). The ticks are thus preferentially attracted to suitable hosts with fed males already attached to them and are not readily attracted to potentially unsuitable hosts without fed males on them (Norval et al. 1989b).

This pheromone-regulated behaviour is of considerable importance in the survival of the ticks. It largely eliminates their chances of attaching to unsuitable hosts, on which they would be unlikely to survive, because of grooming and other factors such as regular treatment with acaricides. For this reason acaricide treatment of cattle appears to have limited effect on population size of *A. hebraeum* if untreated wild hosts, such as giraffe (*Giraffa camelopardalis*), eland (*Taurotragus oryx*) or buffalo, which can carry large numbers of adults, share the same pastures as cattle (Norval and Lawrence 1979). Howell et al. (1981) noted that *A. hebraeum* had not been eradicated from many areas in southern Africa despite intensive acaricide treatment of cattle for half a century. It is reasonable to suppose that pheromone-mediation of host selection was a factor contributing to this.

In considering the effects of the AAP it is worth noting that all stages of *A. hebraeum* can feed repeatedly on the same animals (Norval 1978; Norval et al. 1988b; Holley and Petney 1988). The only manifestation of resistance appears to be increased grooming and this is to some extent countered by the AAP which attracts ticks to areas of the host that are least effectively groomed (Norval et al. 1988b).

Other aspects of the behaviour of *Amblyomma* ticks that are clearly of importance in the epidemiology of heartwater are the behaviour of males on the host and the daily drop-off rhythms of engorged larvae and nymphs.

In both *A. hebraeum* and *A. variegatum* males remain attached to hosts for long periods and mate with numerous females (Jordaan and Baker 1981; Garris 1984). Males of *A. hebraeum* have been shown to remain attached for up to 8 months and to mate as many as 42 times (Jordaan and Baker 1981). Males of both species that are forcibly removed from one host will readily attach to another (Garris 1984; Andrew and Norval 1989a). Males of *A. hebraeum* have also been shown to migrate, on their own accord, from one host to another and from dead to live hosts (Andrew and Norval 1989a). As a result of the long attachment periods, males accumulate on suitable hosts and invariably outnumber females.

Herds or flocks of ruminants infested with *Ambylomma* males will emit larger quantities of CO_2 and AAP than individual animals and so are likely to attract unfed nymphs and adults from greater distances. The practice of herding domestic stock during the day and holding them in enclosures overnight is thus likely to increase the exposure of the animals to unfed ticks. Adults of *A. hebraeum* have been attracted to CO_2/AAP-baited traps from 6 m within 30 min and 11 m within 90 min (Norval et al. 1989a) and are known to seek hosts at night (Norval et al. 1989c).

It is well established in ixodid ticks that the distribution of each stage in the environment is determined primarily by the spatial distribution of hosts, at the time of detachment of engorged ticks of the preceding stage (Minshull and

Norval 1982). In *A. hebraeum* it has been found by Rechav (1978) that most engorged larvae and nymphs detach from hosts towards the end of photophase or shortly after the onset of darkness. This means that the majority of unfed nymphs and adults will occur in and around the areas where hosts congregate at sunset and in the early evening. Where domestic animals are held overnight in enclosures or paddocks, these sites are likely to become the foci of the tick/host interaction. With wild hosts, such foci are likely to be patches of dense vegetation, where birds roost overnight and mammals tend to shelter.

4.4 Infection of Vectors

Until recently it was believed that heartwater infection rates in vector populations were generally low. Uilenberg (1971) reported infection rates of < 1% for adults of *A. variegatum* in Madagascar and Camus and Barre (1987) recorded infection rates of 1–2% for adults of the same species on Guadeloupe. Du Plessis (1985) and Du Plessis and Malan (1987) recorded infection rates of 5–7% in adults of *A. hebraeum* in South Africa.

A problem encountered in all these studies has been the estimation of infection rates by indirect means due to the lack of a direct method for determining the presence or absence of *C. ruminantium* in ticks. Uilenberg (1971) calculated infection rates from tick numbers and the time to infection of domestic ruminants in endemic areas. The other authors inoculated homogenates of single feeding ticks, collected in the field, into mice and then tested the mice for antibodies to *Cowdria* using an immunofluorescent antibody (IFA) test. This method probably underestimates infection rates; two important reasons being the large variation in infectivity of *Cowdria* stocks for mice (Mackenzie and McHardy 1987) and that ticks collected from hosts in the field could have lost their infectivity. There have also been problems in interpreting the results of the IFA test used (Du Plessis et al. 1987).

A different method was used by Norval et al. (1990) in a study on infection rates in unfed nymphs and adults of *A. hebraeum*, collected from two locations in Zimbabwe. Small pools of ticks (± 5) were allowed to feed on sheep known to be susceptible, some of which became infected. The infection rates of the ticks were then estimated statistically. The resulting rates, which varied depending on date of collection and locality, ranged from 0–13% in nymphs, 0–45% in adult males and 20–36% in adult females. Although most infection rates were considerably higher than was previously believed to occur, these could still have been underestimates as the test could have had false negatives.

Infection rates in unfed ticks clearly require further investigation. This will be facilitated by the development of direct methods for detecting *C. ruminantium* in ticks, such as DNA probes.

Cowdria infection rates in the unfed ticks are likely to be affected by several factors. One is the infectivity of the hosts on which the larval and nymphal stages feed. Larvae have a wider host range than nymphs and consequently are likely to feed on more non-carrier hosts. The unfed nymphs, which moult from larvae, are thus less likely to be infected than the unfed adults, which moult from nymphs. Another factor is the apparent difference between larvae and nymphs in their ability to acquire infection from carriers (as previously discussed). There are also differences in the abilities of the different vector species to pick up and transmit *C. ruminantium* (reviewed by Uilenberg 1983a; Bezuidenhout 1987b). For example, *A. hebraeum* and *A. variegatum* can transmit infection from the larval to the nymphal stage, from the nymphal to the adult stage and from the larval through the nymphal to the adult stage, whereas *A. cajennense* and *A. sparsum* have only been shown to transmit from the larval to the nymphal stage. In addition, there is evidence, from studies on *A. variegatum* in Kenya, that susceptibility to *Cowdria* infection may vary considerably between populations of one vector species (Hahn et al. 1990). This last factor could account for the apparent differences in vector infection rates seen in the field (Du Plessis 1982; Camus and Barre 1987) and may help explain observed differences in the epidemiology of heartwater at different locations. Differences in the abilities of vector species to transmit heartwater emphasize the importance of information on vector taxonomy and distribution in studies on the epidemiology of the disease.

Camus and Barre (1982) considered infected, unfed ticks to be the principal reservior of heartwater infection. This view was based on the observations of Ilemobade (1976) that the organism can survive in the vector (*A. variegatum*) for at least 15 months and Neitz (1968) that a generation of *A. hebraeum* may remain infective for up to 3 years. The recent finding of high infection rates in field populations of *A. hebraeum* in Zimbabwe confirms the importance of unfed nymphs and adults as reservoirs of infection.

Another epidemiologically significant discovery has been that adult males of *A. hebraeum* are efficient vectors of heartwater (Andrew and Norval 1989a). The ticks may become infected as nymphs, transmit infection transstadially, move to another host and then transmit intrastadially. They may also become infected as adults and transmit the disease intrastadially. Intrastadial transmission can occur repeatedly and only small numbers of males are required to transmit infection. It can also occur when males migrate from a dead to a live host. These observations suggest that males become highly infected, possibly because of the continuous multiplication of parasites within them, and play an important role in the epidemiology of heartwater as another reservoir of infection.

The importance of attached males, which feed intermittently on one or more hosts during a prolonged parasitic phase, probably extends beyond being simply a source of infection. Repeated infection of individual hosts by accumulations of males could well reinforce the carrier state, as it has been reported that the infectivity of immune hosts increases following re-exposure to heartwater (Neitz et al. 1947). The maintenance of a state of high infectivity would obviously contribute to high infection rates in the unfed ticks.

5 Discussion

The belief, that heartwater was a disease with a short infective period in ruminants and low infection rates in vectors, was not reconcilable with field observations on the epidemiology of the disease, especially the wide occurrence of enzootic stability (Bezuidenhout and Bigalke 1987).

Recent findings on the occurrence of the carrier state in ruminants, high infection rates in vectors and the biology and ecology of vectors, indicate a close and complex interaction between *C. ruminantium*, *Amblyomma* ticks and their vertebrate hosts. The main vectors of heartwater are principally parasites of ruminants, at least in the adult stage. Their distributions in the environment and interactions with ruminants are influenced by clearly defined daily drop-off rhythms in the engorged ticks. The attachment of unfed nymphal and adult ticks to hosts is regulated by the male-produced AAP and this ensures attraction and attachment to suitable hosts, as well as attraction to the areas of the host that are groomed least effectively. It is likely that the AAP also ensures the early exposure of young ruminants to unfed ticks, and hence heartwater, because the ticks will be attracted to the herds or flocks that they are part of and pheromone will be rubbed from dams and others onto the young animals. The attached males, which accumulate on hosts, have several functions; to produce AAP, to mate with feeding females and probably to continuously re-infect hosts with heartwater. The latter function is likely to maintain hosts in an infective state and so contribute to the high vector infection rates that can occur in the field.

There appear to be three important reservoirs of heartwater infection; ruminant carriers, unfed *Amblyomma* nymphs and adults and attached *Amblyomma* males. *Cowdria ruminantium* is therefore an organism that is widely spread through the host and vector populations and this appears to be the reason that enzootic stability for heartwater normally occurs in areas where vectors are well established.

In endemic areas in Africa it is thus more practical to manage heartwater than to attempt to eradicate it. Nevertheless intensive tick control has frequently been and is still, in some circumstances, recommended as the method of choice for heartwater control (Howell et al. 1981; Bezuidenhout and Bigalke 1987).

The first effects of intensive tick control in endemic situations will be the reduction of the sizes of the reservoirs of infection in ticks. This is likely to be followed by reductions in the numbers of animals that become carriers. Rechallenge of existing carriers will also be lowered, possibly reducing the infectivity of these animals, the result of which is likely to be diminished infection rates in ticks. The overall effect will be to reduce the numbers of animals that acquire immunity to heartwater through natural infection. This is the reason why intensive dipping can lead to increased losses due to heartwater (Howell et al. 1981; Bezuidenhout and Bigalke 1987).

Although intensive tick control may lead to the localised eradication of *Amblyomma* ticks, it is unlikely to do so if alternate hosts for the adults are present (due to the effect of the AAP). The importance of wild hosts in maintaining populations of *A. hebraeum* has recently been confirmed by Petney and Horak (1987), who found that on a farm in South Africa the abundance of the tick on wild hosts was reduced to a stable level but thereafter was "apparently unaffected by the level of control" on domestic stock. If the wild hosts on a farm are also heartwater carriers, it will mean that the non-immune domestic stock will be continuously at risk of infection.

The effects of intensive tick control on the epidemiology of heartwater make it a disease that occurs most commonly in east and southern Africa where large numbers of cattle are regularly dipped for the control of *Rhipicephalus appendiculatus*, the main vector of ECF and related diseases. Overall, the heartwater problem appears to be most serious in southern Africa. Uilenberg (1983a) attributes this to the large number of exotic, more susceptible, ruminants present in the subcontinent. While this is undoubtedly of considerable importance another factor may be involved; the vector species. In Sudan, Karrar (1968) reported that heartwater was less of a problem in areas infested with *A. variegatum* than in those infested with other vectors. Similarly, in Zimbabwe, Norval (1983) noted an apparent absence of clinical heartwater in the northwest where the vector is *A. variegatum* and a severe problem in the south where the vector in *A. hebraeum*. The main vector throughout southern Africa is *A. hebraeum*. How the epidemiology of the disease is affected by different vectors is unknown. Factors such as infection rates, seasonal occurrence and host specificity could possibly be involved; further investigation is clearly warranted.

It is only in the western hemisphere, where heartwater and *A. variegatum* have been introduced and are still restricted to limited island foci, that complete vector eradication is recommended (Burridge 1985; Barre et al. 1987). The eradication of *A. variegatum* has already been achieved from Puerto Rico and St. Croix by the treatment of all livestock in affected areas with acaricides (Garris 1987). Its eradication from the remaining islands should be possible as there are few indigenous large ruminants which could serve as alternate hosts for the adults (Barre et al. 1987).

Heartwater management may simply involve the maintenance of enzootic stability by ensuring that livestock are infested with *Amblyomma* ticks. However, if vector numbers are low, because the environment is unsuitable for their survival or because of acaricide use, immunization is the recommended control method. Present methods of immunization involve infection with virulent stocks of *C. ruminantium* and, in those animals not protected by age-related resistance, treatment of the reactions with anti-rickettsial drugs (reviews by Uilenberg 1983a; Oberem and Bezuidenhout 1987b). These are risky procedures and need to be carried out under strict veterinary supervision (Van der Merwe 1979, 1987). To overcome these problems, which impose considerable limitations on the use of immunization, attempts are being made to develop safe, non-living vaccines (Yunker 1988).

In conclusion it is noted that recent research findings have increased, considerably, our understanding of the biological processes involved in the epidemiology of heartwater. This has made it possible to evaluate the likely effects of intensive dipping, a control measure that was widely advocated in the past, and to consider alternate strategies. The approach that is most biologically sound appears to be the management of the disease with or without the use of vaccines. However, before specific recommendations can be made for given situations more quantitative data on the epidemiology of heartwater are required. Obstacles such as lack of diagnostic tests to detect *Cowdria* organisms or antibodies to them should be overcome with the in vitro culture of the organism, improving prospects for further progress in epidemiological studies.

The two major objectives of heartwater research in the future should be the production of improved vaccines and the development of improved diagnostic tests of high sensitivity and specificity. The former to provide a means of control and the latter to allow meaningful estimates to be made of the economic impact of *C. ruminantium* infection and to facilitate the development of epidemiological models for use in the design of cost-effective control strategies.

Acknowledgements. This work was funded by the United States Agency for International Development (Contract No: AFR-0000-C-00-6003-00) through the University of Florida and by the Government of Zimbabwe.

References

Alexander RA, Neitz WO, Adelaar TF (1946) Heartwater. Farming S Afr 21:245, 548–552

Ambrosio RE, Du Plessis JL, Bezuidenhout JD (1987) The construction of genomic libraries of *Cowdria ruminantium* in an expression vector. Onderstepoort J Vet Res 54:255–256

Andrew HR (1990) The spread of heartwater in Zimbabwe in recent years. Farm World 16:11

Andrew HR, Norval RAI (1989a) The role of males of the bont tick *Amblyomma hebraeum* in the transmission of *Cowdria ruminantium.* Vet Parasitol 34:15–23

Andrew HR, Norval RAI (1989b) The carrier status of sheep, cattle and African buffalo recovered from heartwater. Vet Parasitol 34:261–266

Barre N, Camus E (1987) The reservoir status of goats recovered from heartwater. Onderstepoort J Vet Res 54:435–437

Barre N, Uilenberg G, Morel PC, Camus E (1987) Danger of introducing heartwater onto the American mainland: potential role of indigenous and exotic *Amblyomma* ticks. Onderstepoort J Vet Res 54:405–417

Bezuidenhout JD (1987a) Present state of *Cowdria ruminantium* cultivation in cell lines. Onderstepoort J Vet Res 54:205–210

Bezuidenhout JD (1987b) Natural transmission of heartwater. Onderstepoort J Vet Res 54:349–351

Bezuidenhout JD, Bigalke RD (1987) The control of heartwater by means of tick control. Onderstepoort J Vet Res 54:525–528

Bezuidenhout JD, Patterson CL, Barnard JH (1985) In vitro culture of *Cowdria ruminantium.* Onderstepoort J Vet Res 52:113–120

Burridge MJ (1985) Heartwater in the Caribbean. Parasitol Today 1:175–177

Camus E, Barre N (1982) La Cowdriose (heartwater). Revue generale des connaissances. IEMVT, Maisons-Alfort, 147 pp

Camus E, Barre N (1987) Epidemiology of heartwater in Guadeloupe and in the Caribbean. Onderstepoort J Vet Res 54:419–426

Clark R (1962) The pathological physiology of heartwater *Cowdria* (*Rickettsia*) *ruminantium* Cowdry, 1926. Onderstepoort J Vet Res 29:25–33

Du Plessis JL (1982) The effect of intensive tick control on the immunity of cattle to heartwater. In: Proceedings of a symposium on ectoparasites of cattle. South African Bureau of Standards, Pretoria, pp 18–27

Du Plessis JL (1985) A method for determining the *Cowdria ruminantium* infection rate of *Amblyomma hebraeum*: effects in mice infected with tick homogenates. Onderstepoort J Vet Res 52:55–61

Du Plessis JL, Kumm NAL (1971) The passage of *Cowdria ruminantium* in mice. J Sth Afr Vet Med Assoc 42:217–221

Du Plessis JL, Malan L (1987) The application of the indirect fluorescent antibody test in research on heartwater. Onderstepoort J Vet Res 54:319–325

Du Plessis JL, Camus E, Oberem PT, Malan L (1987) Heartwater serology: some problems with the interpretation of results. Onderstepoort J Vet Res 54:327–329

Garris GI (1984) Colonization and life cycle of *Amblyomma variegatum* (Acari: Ixodidae) in the laboratory in Puerto Rico. J Med Entomol 21:86–90
Garris GI (1987) *Amblyomma variegatum* (Acari: Ixodidae): population dynamics and hosts used during an eradication program in Puerto Rico. J Med Entomol 24:82–86
Gruss B (1987) Problems encountered in the control of heartwater in Angora goats. Onderstepoort J Vet Res 54:513–515
Hahn NE, Norval RAI, Waghela SD, Shaw MK, Kocan KM, Rurangariwa F, Perry BD (1990) Different gut infection rates with *Cowdria ruminantium* in two laboratory colonies of *Amblyomma variegatum* ticks Vet Microbiol (in press)
Hess E, de Castro JJ (1986) Field tests of the response of female *Amblyomma variegatum* (Acari: Ixodidae) to the synthetic aggregation-attachment pheromone and its components. Exp Appl Acarol 2:249–255
Holley AD, Petney TN (1988) The use of domestic chickens as laboratory hosts of larvae of the bont tick, *Amblyomma hebraeum*. Onderstepoort J Vet Res 55:75–76
Hoogstraal H (1956) African Ixodoidea. 1. Ticks of the Sudan (with special reference to Equatoria Province and with preliminary reviews of the genera *Boophilus*, *Margaropus* and *Hyalomma*). Research Report NM 005.050.29.07, Department of the Navy, Bureau of Medicine and Surgery, Washington, DC, 1101 pp
Horak IG, Knight MM (1986) A comparison of the tick burdens of wild animals in a nature reserve and on an adjacent farm where tick control is practised. J Sth Afr Vet Assoc 57:199–203
Howell CJ, de Vos AJ, Bezuidenhout JD, Potgieter FT, Barrowman PR (1981) The role of chemical tick eradication in the control or prevention of tick-transmitted diseases of cattle. In: Whitehead GB, Gibson JD (eds) Tick biology and control. Tick Research Unit, Rhodes University, Grahamstown, pp 61–65
Ilemobade AA (1976) Study on heartwater and the causative agent *Cowdria ruminantium* (Cowdry, 1925) in Nigeria. PhD Thesis, Ahmadu Bello University, Zaria, Nigeria, 299 pp
Jongejan F, Wassink LA, Thielemans MJC, Pirie NM, Uilenberg G (1986) Serotypes in *Cowdria ruminantium* and relationship with *Ehrlichia phagocytophila* determined by immunofluorescence. Vet Microbiol 21:31–40
Jordaan JO, Baker JAF (1981) Survival rate on the host and mating capacity of *Amblyomma hebraeum* (Koch) male ticks. In: Whitehead GB, Gibson JD (eds) Tick biology and control. Tick Research Unit, Rhodes University, Grahamstown, pp 115–117
Karrar G (1986) Epizootiological studies on heartwater in the Sudan. Sudan J Vet Sci Anim Husb 9:328–343
Kocan KM, Morzaria SP, Voigt WP, Kairie J, Irvin AD (1987) Demonstration of colonies of *Cowdria ruminantium* in midgut epithelial cells of *Amblyomma variegatum*. Am J Vet Res 48:356–360.
Logan LL, Whyard TC, Quintero JC, Mebus CA (1987) The development of *Cowdria ruminantium* in neutrophils. Onderstepoort J Vet Res 54:197–204
MacKenzie PKI, McHardy N (1987) *Cowdria ruminantium* infection in the mouse: a review. Onderstepoort J Vet Res 54:267–269
MacKenzie PKI, Van Rooyen RE (1981) The isolation and culture of *Cowdria ruminantium* in albino mice. In: Whitehead GB, Gibson JD (eds) Tick biology and control. Tick Research Unit Rhodes University, Grahamstown, pp 33–39
Matheron G, Barre N, Camus E, Gogue J (1987) Genetic resistance of Guadeloupe native goats to heartwater. Onderstepoort J Vet Res 54:337–340
Minshull JI, Norval RAI (1982) Factors influencing the spatial distribution of *Rhipicephalus appendiculatus* in Kyle Recreational Park, Zimbabwe. Sth Afr J Wildl Res 12:118–123
Neitz WO (1935) The blesbuck (*Damaliscus albifrons*) and black wildebeest (*Connochaetes gnu*) as carriers of heartwater. Onderstepoort J Vet Sci Anim Ind 5:35–40
Neitz WO (1968) Heartwater. Bull Off Int Epizoot 70:329–339
Neitz WO, Alexander RA (1941) The immunization of calves against heartwater. J S Afr Vet Med Assoc 12:103–111
Neitz WO, Alexander RA, Adelaar TF (1947) Studies of immunity in heartwater. Onderstepoort J Vet Sci Anim Ind 21:243–249

Norval RAI (1975) Studies on the ecology of *Amblyomma marmoreum* Koch 1844 (Acarina: Ixodidae). J Parasitol 61:737–742
Norval RAI (1977a) Ecology of the tick *Amblyomma hebraeum* Koch in the eastern Cape Province of South Africa. I. Distribution and seasonal activity. J Parasitol 63:734–739
Norval RAI (1977b) Studies on the ecology of the tick *Amblyomma hebraeum* Koch in the eastern Cape Province of South Africa. II. Survival and development. J Parasitol 63:740–747
Norval RAI (1978) Repeated feeding of *Amblyomma hebraeum* (Acarina: Ixodidae) immatures on laboratory hosts. Host effects on tick yield, engorged weight and engorgement period. J Parasitol 64:910–917
Norval RAI (1979) The limiting effect of host availability for the immature stages on population growth in economically important ixodid ticks. J Parasitol 63:285–287
Norval RAI (1983) The ticks of Zimbabwe. VII. The genus *Amblyomma* Zimbabwe Vet J 14:3–18
Norval RAI (1989) Heartwater disease *Cowdria ruminantium* (infection) in domestic ruminants: control and epidemiology. In: Speedy A (ed) Developing world series – agriculture. Grosvenor Press International, London (in press)
Norval RAI, Lawrence JA (1979) The control of heartwater in Zimbabwe Rhodesia. Zimbabwe Rhod Agric J 76:161–165
Norval RAI, Yunker CE, Butler JF (1987) Field sampling of unfed adults of *Amblyomma hebraeum* Koch. Exp Appl Acarol 3:213–217
Norval RAI, Yunker CE, Gibson JD, Deem SLD (1988a) Sampling of unfed nymphs of *Amblyomma hebraeum* Koch. Exp Appl Acarol 4:173–177
Norval RAI, Floyd RB, Kerr JD (1988b) Ability of adults of *Amblyomma hebraeum* (Acarina: Ixodidae) to feed repeatedly on sheep and cattle. Vet Parasitol 29:351–355
Norval RAI, Andrew HR, Yunker CE (1989a) Pheromone-mediation of host selection in bont ticks (*Amblyomma hebraeum* Koch). Science 243:364–365
Norval RAI, Butler JF, Yunker CE (1989b) Use of carbon dioxide and natural or synthetic aggregation-attachment pheromone of the tick, *Amblyomma hebraeum*, to attract and trap unfed adults in the field. Exp Appl Acarol 7:171–180
Norval RAI, Sutherst RW, Jorgensen OG, Gibson JD, Kerr JD (1989c) The effect of the bont tick *Amblyomma hebraeum* on the weight gain of Africander steers. Vet Parasitol 33:329-341
Norval RAI, Andrew HR, Yunker CE (1990) Infection rates with *Cowdria ruminantium* of field populations of nymphs and adults of the bont tick *Amblyomma hebraeum* in Zimbabwe. Vet Parasitol 36:277–283
Norval RAI, Andrew HR, Meltzer MI (1991) Seasonal occurrence of the bont tick (*Amblyomma hebraeum*) in the southern lowveld of Zimbabwe. Exp Appl Acarol 13:81–96
Oberem PT, Bezuidenhout JD (1987a) Heartwater in hosts other than domestic ruminants. Onderstepoort J Vet Res 54:271–275
Oberem PT, Bezuidenhout JD (1987b) The production of heartwater vaccine. Onderstepoort J Vet Res 54:485–488
Pegram RG, Banda DS (1989) Ecology and phenology of cattle ticks in Zambia. Development and survival of free-living stages. Exp Appl Acarol 8:291–301
Pegram RG, Hoogstraal H, Wassef HY (1981) Ticks (Acari: Ixodoidea) of Ethiopia. I. Distribution, ecology and host relationships of species infesting livestock. Bull Entomol Res 71:339–359
Pegram RG, Perry BD, Musisi FL, Mwanaumo B (1986) Ecology and phenology of ticks in Zambia: seasonal dynamics on cattle. Exp Appl Acarol 2:25–45
Pegram RG, Mwase ET, Zivkovic D, Jongejan F (1988) Morphogenetic diapause in *Amblyomma variegatum* (Acari: Ixodidae). Med Vet Entomol 2:301–307
Petney TN, Horak IG (1987) The effects of dipping on parasitic and free-living populations of *Amblyomma hebraeum* on a farm and on an adjacent nature reserve. Onderstepoort J Vet Res 54:529–533
Petney TN, Horak IG, Rechav Y (1987) The ecology of the African vectors of heartwater, with particular reference to *Amblyomma hebraeum* and *Amblyomma variegatum*. Onderstepoort J Vet Res 54:381–395
Provost A, Bezuidenhout JD (1987) The historical background and global importance of heartwater. Onderstepoort J Vet Res 54:165–169

Prozesky L, Du Plessis JL (1984) The pathology of heartwater. A pathological study of sheep, goats and mice infected with *Cowdria ruminantium*. Proc XIII Wld Congr Dis Cattle 1:547–549

Rechav Y (1978) Drop-off rhythms of engorged larvae and nymphs of the bont tick, *Amblyomma hebraeum* (Acari: Ixodidae), and the factors that regulate them. J Med Entomol 14:677–687

Rechav Y (1979) Migration and dispersal patterns of three African ticks (Acari: Ixodidae) under field conditions. J Med Entomol 16:150–163

Rechav Y (1982) Dynamics of tick populations (Acari: Ixodidae) in the Eastern Cape Province of South Africa. J Med Entomol 19:679–700

Rechav Y, Norval RAI, Oliver JH (1982) Interspecific mating of *Amblyomma hebraeum* and *Amblyomma variegatum* (Acari: Ixodidae). J Med Entomol 19:139–142

Thomas AD, Mansvelt PR (1957) Immunization of goats against heartwater. J Sth Afr Vet Med Assoc 28:163–168

Uilenberg G (1971) Etudes sur la cowdriose a Madagascar (Premier partie). Rev Elev Med Vet Pays Trop 24:239–249

Uilenberg G (1981) Heartwater disease. In: Ristic M, McIntyre I (eds) Diseases of cattle in the tropics. Martinus Nijhoff, The Hague, pp 345–360

Uilenberg G (1982) Experimental transmission of *Cowdria ruminantium* by the Gulf Coast tick *Amblyomma maculatum*: danger of introducing heartwater and benign African theileriasis onto the American mainland. Am J Vet Res 43:1279–1282

Uilenberg G (1983a) Heartwater (*Cowdria ruminantium* infection): current status. Adv Vet Sci Comp Med 27:427–480

Uilenberg G (1983b) Acquisitions nouvelles dans la connaissance du role vecteur de tiques du genre *Amblyomma* (Ixodidae). Rev Elev Med Vet Pays Trop 36:61–66

Van de Pypekamp HE, Prozesky L (1987) Heartwater. An overview of the clinical signs, susceptibility and differential diagnosis of the disease in domestic ruminants. Onderstepoort J Vet Res 54:263–266

Van der Merwe L (1979) Field experience with heartwater (*Cowdria ruminantium*) in cattle. J Sth Afr Vet Assoc 50:323–325

Van der Merwe L (1987) The infection and treatment method of vaccination against heartwater. Onderstepoort J Vet Res 54:489–491

Yeoman GH (1968) Field vector studies of epizootic East Coast Fever. VI. The occurrence of *Amblyomma variegatum* and *A. lepidum* in the East Coast fever zones. Bull Epizoot Dis Afr 16:183–203

Yunker CE (1988) Current research on heartwater in Zimbabwe. Zimbabwe Vet J 19:25–31

Yunker CE, Byrom B, Semu S (1988) Cultivation of *Cowdria ruminantium* in bovine vascular endothelial cells. Kenya Vet 12:12–16

Walker JB, Olwage A (1987) The tick vectors of *Cowdria ruminantium* (Ixodoidea, Ixodidae, genus *Amblyomma*) and their distribution. Onderstepoort J Vet Res 54:353–379

Walker JB, Schulz KCA (1984) Records of the bont tick *Amblyomma hebraeum*, from the angulate tortoise, *Chersina angulata*, and the leopard tortoise, *Geochelone pardalis*. Onderstepoort J Vet Res 51:171–173

Studies of the Role of *Amblyomma variegatum* in the Transmission of *Dermatophilus congolensis*

D. MARTINEZ, N. BARRE, B. MARI and T. VIDALENC[1]

Summary

The role of *Amblyomma variegatum* in the transmission of *Dermatophilus congolensis* was investigated by feeding immature and adult stages of the tick on the skin of goats infected with a culture of a rifampicin resistant strain of *D. congolensis* (R2 strain). Subsequent isolations of the pathogen on selective media showed that 36 and 43% of males carried *D. congolensis* for at least 30 days. There was no evidence that larvae and nymphae carried *D. congolensis*. Transstadial transmission did not occur, indicating that *A. variegatum* has little role if any in the transmission of the pathogen.

The possibility of transmission by direct contact or by insects was proved by demonstrating the development of antibodies to *D. congolensis* in the sera of 65% of naive seronegative goats maintained under tick free conditions and either allowed to mix with other carrier goats or kept isolated in individual cages. The existence of asymptomatic carriers was confirmed by the observation that none of these goats developed lesions in spite of producing high levels of specific antibodies.

Severe dermatophilosis was reproduced by feeding of adult *A. variegatum* on asymptomatic carrier goats, while scarification of the same group of goats with *D. congolensis* resulted in a mild manifestation of the disease. The possible reasons for these observations are discussed in the light of the role played by adult *A. variegatum*.

1 Introduction

Dermatophilosis is an exudative skin disease affecting numerous animal species and man (Stewart 1972a) and is caused by an actinomycete, *Dermatophilus congolensis* (Van Saceghem 1915). The disease, which occurs in most parts of the world is of little economic significance in temperate climates where it affects mainly sheep (Stewart 1972a; Hyslop 1980). However, severe economic losses occur in cattle, small ruminants and horses in the tropics (Ford et al. 1974; Lloyd 1976). Severe dermatophilosis has never been readily reproduced experimentally as the organism per se is not very pathogenic. The appearance of the disease is the result of a complex interplay between environmental and intrinsic factors among which the breed, the climate and the presence of ticks are of primary importance.

In Nigeria, the prevalence of dermatophilosis has been estimated to vary in local cattle from an average of 3–5% in the dry season to 10–12% in the wet season (Bida 1975; Oduye 1975a). In Madagascar, Ranaivoson et al. (1986) observed that less than 1% of local zebu cattle were affected by dermatophilosis

[1] Institut d'Elvage et de Médécine Vétérinaire des Pays Tropicaux, BP 1232, 97148, Pointe a Pitre Cedex, Guadeloupe, French West Indies

compared with 29 to 85% of exotic cattle among which a mortality rate of 67% was reported. These results are consistent with those of other reports from Africa (Balabanov and Boussafou 1977; Leroy and Marchot 1987), Madagascar (Blancou 1976) and the Caribbean (Barré et al. 1988).

Whatever the influence of breed and season, it has long been noted that outbreaks are associated with the presence of the ticks, *Amblyomma variegatum* (Plowright 1956; Norval 1986).

A. variegatum was introduced into the Caribbean islands in the last century following importation of cattle from Africa (Curasson 1943). The disease is of economic importance only in islands infected with *A. variegatum*. Acute dermatophilosis is restricted to those areas of Martinique and St. Lucia where *A. variegatum* is known to occur (Morel 1967; Butler 1975; Burridge et al. 1984; Garris and Scotland 1985; Morrow et al. 1989; Matheron et al. 1990). Clinical disease is absent or sporadic in islands free of *A. variegatum*. Tick control is currently the most efficient method of controlling and preventing the spread of dermatophilosis (Plowright 1956; Butler 1975; Matheron et al. 1990).

The epidemiology and pathogenesis of dermatophilosis remains poorly understood despite its wide distribution and economic impact on animal production (Lloyd 1984). This paper documents the results of experiments conducted in Guadeloupe on aspects of the role of *A. variegatum* in the epidemiology of dermatophilosis.

2 Materials and Methods

2.1 Experimental Animals and Ticks

A local Guadeloupean strain of *A. variegatum* was reared in the laboratory on goats and maintained at 25 °C and 90% RH for several generations. Goats used for these experiments were imported from Les Saintes islands which are free of dermatophilosis and *A. variegatum*. All studies were conducted on our experimental farm "Duclos" in Guadeloupe, French West Indies.

2.2 Bacterial Strain

A field strain of *D. congolensis* was isolated from clinically infested Creole cattle in Guadeloupe. Rifampicin resistant mutants of this strain were selected in vitro as previously described (Martinez and Prior 1990) and one of them, R2, was used in the experiments. Unlike field strains, such mutants can be isolated on selective medium and can be differentiated from field strains using antibiotic sensitivity as a marker.

2.3 Media

Isolations were performed on Muller-Hinton agar or in selective media consisting of nutrient broth supplemented with 10% neonatal calf serum,

100 μg ml^{-1} rifampicin, 100 μg ml^{-1} nalidixic acid and 100 μg ml^{-1} amphotericin B. Cultures were incubated at 37 °C for 5 days and bacterial counts were carried out on Muller-Hinton agar using the spread plate method.

2.4 Serological Assay

Antibodies to *D. congolensis* were detected by enzyme-linked immunosorbent assay (ELISA) (Martinez et al. 1989). Briefly, polystyrene microplates (Nunc immuno-modules) were coated by passive adsorption with antigens comprising 2 μg ml^{-1} bacterial protein diluted in carbonate-bicarbonate buffer 0.1 M, pH 9.5 (100 μl per well), overnight at 37 °C. After coating, the plates were washed three times in PBS-tween (PBS supplemented with 0.1% tween 20, solution used for all washings). To each well was added 100 μl of goat serum diluted 1/800 in the blocking buffer (PBS tween to which was added 5% skimmed cows milk). The plates were incubated for 1 h at 37 °C. After five washings, 100 μl of rabbit anti-goat IgG conjugated to horse radish peroxidase (SIGMA) and optimally diluted 1/6000 in the blocking buffer was added. The plates were incubated for 1 h at 37°C, washed five times and each well filled with 100 μl of the substrate containing 0.5 mg ml^{-1} O-phenylene diamine and 3 μl ml^{-1} of 9% H_2O_2 in citrate-citric acid buffer 0.1 M, pH 5.5. The enzymatic reaction was stopped after 30 min of incubation at room temperature, by adding 50 μl of 2N H_2SO_4 per well and the absorbance was read at 495 nm in a spectrophotometer.

2.5 Infection of Instars of *A. variegatum* with *D. congolensis*

Infection of *A. variegatum* was carried out by shaving an area of 100 to 150 cm^2 on the skin of goats. This was followed by swabbing 5 ml of a broth culture containing approximately 10^7 cfu (colony forming units) ml^{-1} of the R2 strain of *D. congolensis*. Ticks were applied to the swabbed area in bags glued to the skin with adhesive (nymphae and adults) or in ear bags (larvae). Females were applied 5 days after males had attached. Larvae, nymphae and females were collected after detachment. The remaining males were removed manually. In each experiment the same groups of ticks were recovered in the same manner after feeding on uninfected control goats. Half of the recovered tick instars were retained for bacteriological studies while the remainder were placed in culture at 25 °C, 95% RH and allowed to moult.

2.6 Isolation of *D. congolensis* from *A. variegatum*

Tick instars which had fed on *D. congolensis* infected and uninfected control goats were individually cleaned by washing three times in sterile distilled water (SDW). Nymphae and adult ticks were sectioned behind the first pair of legs and the anterior part of their bodies were placed in individual vials filled with broth selective medium. Isolations were attempted from larvae which were placed whole in the medium. The vials were examined daily for 10 days and cultures microscopically positive for *D. congolensis* were subcultured on selective agar medium.

D. congolensis was isolated from skin lesions by washing dried skin exudate with SDW and plating on selective and non-selective agar medium.

Attempts were made to isolate *D. congolensis* from male *A. variegatum* which had been manually removed from artificially infected and uninfected goats. Isolations were attempted from the following:

1. Thirty males removed from infected and control goats collected 30 days after attachment to the host.
2. A separate batch of 22 males from infected and control goats collected 22 days after attachment, half of which were disinfected for 1 min in 70% ethanol, immersed in 1% solution of sodium hypochlorite for 1 min followed by three washings in SDW.

In a separate study isolations were attempted from recently moulted nymphae and adult males which had previously fed on *D. congolensis* infected and uninfected control goats as described earlier.

2.7 Transmission of *Dermatophilus congolensis* Without Ticks

Twenty goats were imported from Les Saintes, placed immediately in isolation and screened for the presence of antibodies to *D. congolensis* by ELISA. During the experiment which lasted 10 weeks, half were maintained in a 25 m^2 tick-free paddock and mixed with healthy goats which had been in contact with *D. congolensis*, while the other half were placed in separate individual cages which prevented any direct contact with other animals. Sera were collected weekly from the day of their arrival and tested for the development of antibodies to *D. congolensis*.

2.8 The Influence of *A. variegatum* on the Development of Severe Generalized Dermatophilosis

Thirteen goats from Les Saintes were maintained for 10 weeks on the experimental farm after which they were divided into three groups. Two groups of five goats each were infested with five and ten pairs of adult *A. variegatum* respectively while a control group comprising three goats remained tick free but were scarified with surgical blades followed by an application of a culture of R2 strain of *D. congolensis*. Seven weeks later the two groups of goats which had been previously infected with five and ten pairs of *A. variegatum* were again reinfested with the same number of ticks immediately after the attachment sites were infected with R2 strain of *D. congolensis* as described earlier. The control group was again scarified with the R2 strain of *D. congolensis*.

The severity of the skin lesions was scored 1 to 3 and recorded for each of five sites on the body, viz. ears, lips, eyelids, legs and body. The scores for each site were totalled and expressed as a total damage score. The goats were bled weekly and the sera were assayed for *D. congolensis* specific titres.

3 Results

3.1 Isolation of *D. congolensis* from Instars of *A. variegatum* Fed on Infected Skin

D. congolensis was isolated from 13/30 and 8/22 males fed on infected skin, 30 and 22 days respectively after attachment. Similarly, the isolation of *D. congolensis* was successful in 4/22 disinfected and 4/22 non-disinfected adult ticks. There were no isolations of *D. congolensis* from larvae, newly moulted nymphae, adults and ticks fed on control goats.

3.2 Transmission Without Ticks

Thirty percent of seronegative goats from Les Saintes developed antibodies to *D. congolensis* using ELISA within 2 weeks of their arrival in Guadeloupe in spite of being housed under tick-free conditions (Fig. 1). After 10 weeks, 65% of goats which were either isolated in individual cages or mixed with goats had seroconverted. None of the seropositive goats developed any visible lesions of dermatophilosis during the course of the experiment.

3.3 The Influence of *A. variegatum* on the Development of Generalized Dermatophilosis

The results of this study are illustrated in Fig. 2.

Six of the 13 seronegative goats became seropositive to *D. congolensis* without showing clinical symptoms within 10 weeks of their arrival on the

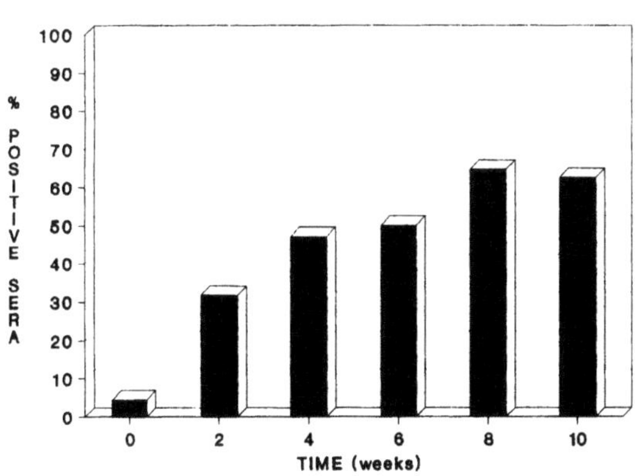

Fig. 1. Percentage of goats from Les Saintes which developed antibodies to *D. congolensis* in tick-free conditions after their introduction onto an infected farm

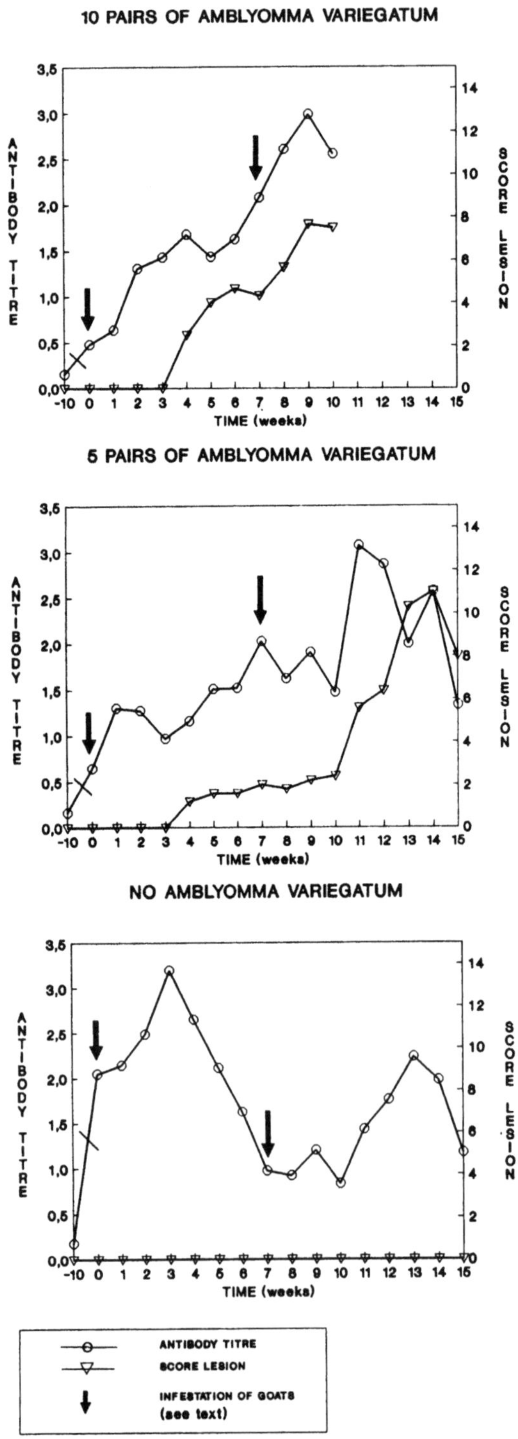

Fig. 2. Antibody titres and dermatophilosis score lesions of goats infested or not with *A. variegatum*, with or without *D. congolensis*

Table 1. The development and course of clinical dermatophilosis on three groups of goats following infestation by ticks and artificial infection with *D. congolensis*

Goat No.	Level of infestation	Time to appearance of lesions (weeks)	Course of disease (weeks)	Survival time (weeks)
32	10 M/10 F[a]	4	6	10
42	"	4	6	10
46	"	4	1	5
50	"	0	0	4
31	"	4	7	11
	Means ± SD:	3.2 ± 1.8	4 ± 3.2	8 ± 3.2
24	5 M/5 F	7	8	15
34	"	4	11	15
43	"	4	9	13
38	"	4	9	13
29	"	4	8	12
	Mean ± SD:	4.6 ± 1.3	9 ± 1.2	13.6 ± 1.3
25	*D. congolensis*	No lesions	No disease	No deaths
40	"	"	"	"
28	"	"	"	"

[a] Male (M) and female (F) *A. variegatum*.

experimental farm. Typical lesions developed after the first infestation of adult ticks only (without *D. congolensis*) and before the second infestation with adult ticks and infection with R2 (Fig. 2). Simultaneous isolation on selective and non-selective media showed that the causative strain of *D. congolensis* was not rifampicin resistant. The second infestation with ticks and R2, induced localized lesions from which R2 could be isolated, but the spread of the disease on the body was due to infection by the field strain of *D. congolensis.*

Except for one goat, which died after 4 weeks before developing lesions, and one, which only developed lesions after 7 weeks, the time to the development of visible lesions after the first tick infestation was very similar in all goats (4 weeks). The period between the development of the lesions and death was significantly reduced when the number of ticks increased from five to ten pairs per animal ($P = 0.01$; Table 1). All goats produced antibodies to *D. congolensis* before developing lesions, while uninfested controls seroconverted without developing overt lesions. The antibody titres roughly followed the lesion score (Fig. 2).

4 Discussion

D. congolensis could be isolated from 43 and 36% of male *A. variegatum* fed on infected goat skin in two experiments. These percentages were higher than those found by Oppong (1976) who isolated *D. congolensis* from 30 and 4% of *A. variegatum* collected from infected and healthy cattle respectively. The use of the rifampicin resistant R2 strain of *D. congolensis* improved the sensitivity of the method by reducing contamination. The isolation from 4 out of the 22 disinfected ticks indicated that the pathogen was probably localized in or around the mouthparts or the anterior part of the digestive tract. It is clear that these ticks remained carriers of *D. congolensis* for at least 1 month. However, male *A. variegatum* do not detach after feeding and remain on their host for long periods (Barré 1989) although some males are capable of transferring to another host. Andrew and Norval (1989) observed movement of male *Amblyomma hebraeum* on the host and transfer from a dead to a live animal resulting in the transmission of heartwater. The extent to which this occurs in *A. variegatum* is unknown and is probably an uncommon occurrence. The role of males in the transmission of *D. congolensis* is therefore likely to be limited. *D. congolensis* could not be isolated from larvae or nymphae which had engorged on infected skin, nor from nymphae and adult males after moulting. Although this observation needs to be repeated by a more sensitive method such as ELISA, it would appear that no transstadial transmission of *D. congolensis* occurs. Macadam (1962) succeeded in transmitting the disease in one case to a rabbit with *A. variegatum* collected on infected cattle, but the developmental stage of the collected ticks was not specified, and it is not clear if the transmission was transstadial or direct from the harvested ticks. Kusel'tan (1967) cited by Stewart (1972b) who recovered the organism from *Hyalomma asiaticum* also claimed that the tick transmitted the disease, but he did not demonstrate the existence of a transstadial transmission of *D. congolensis*, an important condition for a tick to play a role in the transmission of dermatophilosis. It would appear that *A. variegatum* seems to play an insignificant role if any, as a biological vector of *D. congolensis*.

Transmission was possible in the absence of ticks. In effect, 65% of goats kept under tick free conditions developed antibodies to *D. congolensis* within 10 weeks of arrival from a tick-free island. Transmission could have occurred by direct contact with contaminated material and animals, a method which is regarded as important by Oppong (1976) and Hyslop (1980), although it is questioned by Bull (1929) and Plowright (1956).

The transmission of *D. congolensis* to goats which had been isolated in individual cages since their arrival, could have been caused by insect bites. The high rate of transmission in the present trial is in agreement with the results of Richard and Pier (1966) who have demonstrated experimentally the transfer of infection to recipient rabbits by *Musca* spp. and by *Stomoxys calcitrans* in 30 out of 45 transmission trials up to 24 h after the flies were permitted to feed on donor rabbits. By contrast, Philpott and Ezeh (1978) were able to induce lesions on cattle in only one case with *S. calcitrans* and one case with *Musca domestica*. These dissimilar results may be due to differences in susceptibility of rabbits and cattle, although the observation of lesions alone is not a good criterion for confirmation of transmission. Esterre and Agis (1983) reported the isolation of the bacterium from the legs of *Cochlioma macellaria*. The organism was observed once on Giemsa staining of ground mouthparts of *S. calcitrans* by

Philpott and Ezeh (1978) and was isolated in only one group of *M. domestica* by Richard and Pier (1966) in spite of a high transmission rate of the disease. It would therefore appear that bacteriology alone is not a reliable indicator of the role of insects in the transmission of infection. Biting flies have been claimed to play a more important role in the field than ticks by Macadam (1962, 1976). Stephen and Broomfield (cited by Macadam 1964) observed dermatophilosis lesions on the site of challenge with *Glossina* spp. on experimental cattle in Nigeria, and an increased prevalence of the disease in years when cattle were subjected to heavy attacks by Tabanids. *D. congolensis* was observed microscopically in early lesions on the back, flanks and rump of cattle, produced by bites of *Haematopota albihirta* and *Tabanus taeniola* in Zimbabwe at the start of the rainy season. Ticks seldom attached to these particular regions on the host (Stewart 1972b). Calderbrank and Wilson (cited by Macadam 1962) recorded the occurrence of lesions behind the hump due to the bite of *Lyperosia*, and reported a reduction of the incidence of dermatophilosis by control of the flies. Seasonal increases in incidence of disease were noted to coincide with seasonal increases in flies and mosquito activity (Roberts and Graham 1966; Roberts 1967). However, in all these cases, it was not certain whether insects transferred the zoospores or provided a skin injury inducing a primary lesion. The transmission by *Musca* spp. proves that an important mechanical disruption of the host's skin is not a necessary prerequisite for the establishment of infection (Richard and Pier 1966).

It is likely that other undetermined factors may influence the epidemiology of dermatophilosis and it is possible that estimates of infection rates in carriers may be low. Serology is presently the most reliable method of monitoring infection rates in animal populations especially at the subclinical level. The role of mosquitoes, ceratopogonids, *Stomoxys* and *Haematobia* which are abundant in Guadeloupe requires investigation.

The study to demonstrate the role of *A. variegatum* in the pathogenesis of dermatophilosis is of interest. A severe form of the disease developed in the goats despite the high titres of *D. congolensis* specific antibodies. Furthermore, the disease progressed despite the detachment of the ticks. The isolation of the field strain of *D. congolensis* instead of laboratory R2 strain demonstrated the role of naturally infected carriers within the host population. Six of 13 goats had already developed high antibody titres without scabs before the start of the experiment and it is likely that the remainder were also infected by the field strain which had not yet induced a detectable antibody response. It is interesting to note that R2 provoked only localised lesions after the second infestation, while the field strain induced severe dermatophilosis. These observations may be due to differences in pathogenicity between the field and laboratory strain or that the R2 strain was superimposed on an already established wild population on the host. It is clear that the expression of the disease was influenced by infestation with *A. variegatum* since none of the goats present on the experimental farm where the wild strain exists, developed dermatophilosis.

It would appear that a simple disruption of the epidermis alone is not sufficient to induce a severe form of dermatophilosis contrary to the opinion of Thiery and Memery (1961) and Zlotnik (1955), since in this case the same mechanical disruption produced by scarification of controls with R2 should have provoked the disease. Oduye (1975b) was also unable to reproduce a lesion by pricking the skin with an infected needle. The close association between extensive dermatophilosis and *A. variegatum* and not the closely related species *A. hebraeum* (Norval 1986), suggests that the role of the former is not simply to provide a portal of entry for the organism.

Although experimental abrasion of the skin by mechanical or chemical means has been shown to favour the development of lesions (Oduye 1975b; Lloyd and Jenkinson 1980; Lloyd and Noble 1982), the natural disease has never been reproduced by such treatment. It is clear that more complex mechanisms influence the pathogenesis of dermatophilosis. Poxviruses have been found to be associated with *D. congolensis* in the same lesion and thereby incriminated as a synergistic pathogen (Munz 1976; Abu-Samra 1978; Abu-Samra and Walton 1981; Isitor et al. 1988). Reproduction of severe dermatophilosis has never been demonstrated with poxvirus in the absence of *A. variegatum.*

Uilenberg (1987) failed to demonstrate a correlation between the presence of antibodies to *Theileria mutans* or *T. velifera* transmitted by *A. variegatum* and lesions of dermatophilosis. The positive response to treatment with long acting tetracycline suggests that if such a concomitant infectious agent does exist then it is sensitive to tetracycline and its destruction is sufficient to arrest the development of the lesion. Arthropod saliva is known to contain numerous biologically active substances (Ribeiro 1987) which could influence the development of the lesions. The crusts experimentally produced on cattle by feeding ticks on an area rubbed with a *Dermatophilus* culture were shown to be larger and to persist longer than lesions produced by scarification (Macadam 1962). The same observations were made on goats infected with R2 (Martinez et al. unpubl.).

Davis and Philpott (1980) have shown that a chronic skin lesion of dermatophilosis can be induced in the goat and fail to heal as long as a delayed hypersensitivity reaction is maintained by local application of dinitrochlorobenzene. These authors suggested that such a reaction mimics the immune response of the host at the site of arthropod bites which favours the persistence of the pathogen. This hypothesis is supported by the observations of some authors who report that lesions occur predominantly at the predilection sites of adult *A. variegatum* (Plowright 1956; Macadam 1964) and by the fact that such infection can be eliminated by tick control (Plowright 1956; Butler 1975; Blancou 1976; Matheron et al. 1990). However, this theory does not satisfactorily account for the development of lesions in areas of skin where ticks seldom bite. In the French West Indies, Barré et al. (1988) found that 90% of infected cattle had lesions on the back where less than 1% of ticks attach, an observation which is in accordance with the findings of Morrow et al. (1989) in Antigua.

It has been suggested that the occurrence of dermatophilosis is related to an immune deficiency of the host (Roberts 1967; Davis 1984; Lloyd 1984). Immunosuppressive activity of tick saliva has been demonstrated in *Ixodes dammini* (Ribeiro et al. 1985), in *Dermacentor andersoni* (Wikel and Osborn 1982) and *Rhipicephalus appendiculatus* (Fivaz 1989). This action could favour the transmission of pathogens (Wikel and Whelen 1986) and explain the absence of natural immunity to *A. variegatum* and *A. hebraeum* (Norval et al. 1988; Jongejan et al. 1989). There is a need to investigate the possible immunosuppressive activity of the saliva of *A. variegatum* especially in terms of the effect of the tick on the various populations and subpopulations of cells of the immune system. Wikel and Whelen (1986) have demonstrated that in vitro responsiveness of lymph node cells of Guinea pigs to T-lymphocyte mitogens was significantly reduced by infestation of hosts with *Dermacentor andersoni* larvae, while responsiveness to B-lymphocyte mitogens was unimpaired.

Acknowledgements. Part of this research was supported by the European Community (Directorate General XII) under contract No. TSD-A-249.

References

Abu-Samra MT (1978) Dermatophilosis infection: the clinical disease and diagnosis. Zentralbl Veterinärmed B 25:641–651

Abu-Samra MT, Walton GS (1981) The inoculation of rabbits with *Dermatophilosis congolensis* and the simultaneous infection of sheep with *D. congolensis* and orf virus. J Comp Pathol 91:371–329

Andrew RA, Norval RAI (1989) The role of males of the Bont tick (*Amblyomma hebraeum*) in the transmission of *Cowdria ruminantium* (heartwater). Vet Parasitol 34:15–23

Balabanov VA, Boussafou D (1977) Dermatophilose du betail en Republique Populaire du Congo. Rev Elev Med Vet Pays Trop 30:363–368

Barré N (1989) Biologie et ecologie de la tique *Amblyomma variegatum*, en Guadeloupe (Antilles Francaises). These Doctorat des Sciences, Universite Paris Sud, pp 267

Barré N, Matheron G, Rogez B, Roger F, Martinez D, Sheikboudou C (1988) La dermatophilose des bovins a *Dermatophilus congolensis* dans les Antilles francaises. II. Facteurs de receptivite lies aux animaux. Rev Elev Med Vet Pays Trop 41:339–345

Bida SA (1975) Seasonal prevalence of bovine dermatophilosis in Northern Nigeria. Bull Off Int Epizoot 83:1131–1138

Blancou J (1976) Bilan de sept annees de prophylaxie de la dermatophilose dans un troupeau de zebus brahman. Rev Elev Med Vet Pays Trop 29:211–215

Bull LB (1929) Dermatomycosis of the sheep (lumpy or matted wool) due to *Actinomyces dermatonomus*. Aust J Exp Biol Med Sci 6:301–314

Burridge MJ, Barre N, Birnie EF, Camus E. Uilenberg G (1984) Epidemiological studies on heartwater in the Caribbean, with observations on tick-associated bovine dermatophilosis. Proc XIIIth World Congr Diseases of Cattle, Durban, South Africa 1:542–546

Butler MC (1975) Dermatophilosis of cattle, sheep and goats in St. Lucia. State Vet J 30:279–283

Curasson G (1943) Traite de protozoolgie humaine et comparee. Vigot, Paris, 445 pp

Davis D (1984) Infection with *Dermatophilosis congolensis* at a contact hypersensitivity site and its relevance to chronic streptothricosis lesions in the cattle of west Africa. J Comp Pathol 44:25–32

Davis D, Philpott M (1980) Experimental chronic dermatophilosis. Proc R Soc Edinb 79:47–53·

Esterre P, Agis F (1983) La dermatophilose aux Antilles francaises. Rev Elev Med Vet Pays Trop 36:137–140

Fivaz BH (1989) Immune suppression induced by the brown ear tick *Rhipicephalus appendiculatus* (Neumann) 1901. J Parasitol 75:946–952

Ford RB, Cairns RA, Short CD (1974) Equine dermatophilosis: a two-year clinico-pathologic study. Vet Med Small Anim Clin 69:1557–1561

Garris GI, Scotland K (1985) Ticks on livestock in St. Lucia. Vet Parasitol 18:367–373

Hyslop NStG (1980) Dermatophilosis (streptothricosis) in animals and man. Comp Immunol Microbiol Infect Dis 2:389–404

Isitor GN, Kazeem HM, Njoku CO, Adegboye DS, Dellman HD (1988) Frequency of involvement of pox virions in lesions of bovine dermatophilosis. Trop Anim Health Prod 20:2–10

Jongejan F, Pegram RG, Zivkovic D, Hensen EJ, Mwase ET, Thielmans MJC, Cosse A, Niewold TA, Ashraf ES, Uilenberg G (1989) Monitoring of naturally acquired and artificially induced immunity to *Amblyomma variegatum* and *Rhipicephalus appendiculatus* ticks under field and laboratory conditions. Exp Appl Acarol 7:181–199

Leroy P, Marchot P (1987) The resistance to dermatophilosis of dinka cattle breed, dinka crossbred and boran, friesian, jersey, sahiwal crossbreds. Ann Rech Vet 18:107–109

Lloyd DH (1976) The economic effects of bovine streptothricosis. In: Lloyd DH, Sellers KC (eds) *Dermatophilus* infection in animals and man. Academic Press, London, pp 274–297

Lloyd DH (1984) Immunology of dermatophilosis: recent developments and prospects for control. Prev Vet Med 2:93–102

Lloyd DH, Jenkinson McE (1980) The effect of climate on experimental infection of bovine skin with *Dermatophilus congolensis*. Br Vet J 136:122–134

Lloyd DH, Noble WC (1982) *Dermatophilus congolensis* as a model pathogen in mice for the investigation of factors influencing skin infection. Br Vet J 138:51–60

Macadam I (1962) Bovine streptothricosis: production of lesions by the bites of the tick *Amblyomma variegatum*. Vet Rec 74:643–646

Macadam I (1964) Observations on the effects of flies and humidity on the natural lesions of streptothricosis. Vet Rec 76:194–198

Macadam I (1976) Some observations on *Dermatophilus* infection in the Gambia with particular reference to the disease in sheep. In: Lloyd DH, Sellers KC (eds) *Dermatophilus* infection in animals and man. Academic Press, London, pp 33–40

Martinez D, Prior P (1990) Survival of *Dermatophilus congolensis* in tropical clay soils submitted to different water potentials. Vet Microbiol (in press)

Martinez D, Mari B, Aumont G, Vidalenc T (1989) Detection of IgG antibodies to *Dermatophilus congolensis* in ruminant sera by ELISA. Intl Conf Application of biotechnology to livestock in developing countries. Centre for Tropical Veterinary Medicine, University of Edinburgh, 4–8 Sept 1989

Matheron G, Barré N, Roger F, Rogez B, Martinez D, Sheikboudou C (1990) La *Dermatophilose* des bovins a *Dermatophilus congolensis* dans les Antille francaises. III. Comparisons entre elvages infectes et elvages indemnes. Rev Elev Med Vet Pays Trop 42:331–347

Morel PC (1967) Etude sur les tiques du betail en Guadeloupe et Martinique II. Agents pathogenes transmis par les tiques. Rev Elev Med Vet Pays Trop 20:291–299

Morrow AN, Heron ID, Walker AR, Robinson JL (1989) *Amblyomma variegatum* ticks and the occurrence of bovine streptothricosis in Antigua. J Vet Med 36:241–249

Munz E (1976) Double infection of sheep and goats in Kenya with orf virus and dermatophilus. In: Lloyd DH, Sellers KC (eds) *Dermatophilus* infection in animals and man. Academic Press, London, pp 57–67

Norval RAI (1986) *Amblyomma variegatum* and acute bovine dermatophilosis. Parasitol Today 2:23

Norval RAI, Floyd RB, Kerr JD (1988) Ability of adult *Amblyomma hebraeum* (Acarina: Ixodidae) to feed repeatedly on sheep and cattle. Vet Parasitol 29:351–355

Oduye OO (1975a) Incidence, economic effect and control of bovine streptothriosis in Nigeria. Bull Off Int Epizoot 83:1125–1129

Oduye OO (1975b) Effects of various induced local environmental conditions and histopathological studies in experimental *Dermatophilus congolensis* infection on the bovine skin. Res Vet Sci 19:245–252

Oppong ENW (1976) Epizootiology of *Dermatophilus* infection in cattle in the Accra plains of Ghana. In: Lloyd DH, Sellers KC (eds) *Dermatophilus* infection in animals and man. Academic Press, London, pp 17–32

Philpott M, Ezeh AO (1978) The experimental transmission by *Musca* and *Stomoxys* species of *D. congoloensis* infection between cattle. Br Vet J 134:515–520

Plowright W (1956) Cutaneous streptothricosis of cattle: I. Introduction and epizootiological features in Nigeria. Vet Rec 68:350–355

Ranaivoson A, Ranaivoson R, Rambelmonana D (1986) Epizootiologie et incidence de la dermatophilose bovine a madagascar. Rev Elev Med Vet Pays Trop 39:279–287

Riberio JMC (1987) Role of saliva in blood-feeding by arthropods. Annu Rev Entomol 32:463–478

Riberio JM, Makoul GT, Levine J, Robinson DR, Spielman A (1985) Antihemostatic, antiinflam-

matory, and immunosuppressive properties of the saliva of a tick, *Ixodes dammini*. J Exp Med 161:332–344
Richard JL, Pier AC (1966) Transmission of *Dermatophilus congolensis* by *Stomoxys calcitrans* and *Musca domestica*. Am J Vet Res 27:419–423
Roberts DS (1967) *Dermatophilus* infection. Vet Bull 37:514–521
Roberts DS, Graham PH (1966) Control of ovine cutaneous actinomycosis. Aust Vet J 42:74–78
Stewart GH (1972a) Dermatophilosis: a skin disease of animals and man. Part I. Vet Rec 91:537–544
Stewart GH (1972b) Dermatophilosis: a skin disease of animals and man, Part II. Vet Rec 91:555–561
Thiery G, Memery G (1961) La streptothricose cutanee. IV. Etiologie-traitement-prophylaxie. Rev Elev Med Vet Pays Trop 14:413–427
Uilenberg G (1987) Control of livestock diseases associated with the tick *Amblyomma variegatum*. Final Rep, Fac Vet Med Univ Utrecht, Contract EEC, TSD-A-249, 39 pp
Van Saceghem R (1915) Dermatose contagieuse (Impetigo contagieux). Bull Soc Path Exot 8:354–359
Wikel SK, Osburn RL (1982) Immune responsiveness of the bovine host to repeated low-level infestations with *Dermacentor andersoni*. Ann Trop Med Parasitol 76:405–414
Wikel SK, Whelen AC (1986) Ixodid-host immune interaction. Identification and characterization of relevant antigens and tick-induced host immunosuppression. Vet Parasitol 20:149–174
Zlotnik I (1955) Cutaneous streptothricosis in cattle. Vet Rec 67:613–614

Bovine Ehrlichiosis

C.G. Stewart[1]

Summary

Ehrlichia bovis infection in cattle is seldom reported and is usually associated with subclinical infection. A severe disease which was associated with infection with *Ehrlichia bovis* and known as Nofel, has been reported from the Sudan and other countries in Central Africa. Bovine ehrlichiosis is characterized by irregular fever, drooping of the ears, turning and lymphadenitis. There is a high mortality with animals dying within a few hours in the peracute form of the disease and within 36 to 48 h in the acute form of the disease.

Transmission occurs following bites by infected ticks. *Amblyomma variegatum* and *Rhipicephalus appendiculatus* have been proven to be vectors.

1 Introduction

Ehrlichia are members of the tribe Ehrlichieae in the family Rickettsiaceae, order Rickettsiales (Weiss and Moulder 1984). *Ehrlichia bovis* was first described by Donatien and Lestogard in 1936 at the Algerian Pasteur Institute while conducting transmission studies on *Theileria*. They named the parasite, which occurred in monocytes, *Rickettsia bovis*. Apart from the enlargement of lymphnodes, development of a temperature reaction, and some loss of condition, the parasite was not considered to be pathogenic. Subsequently the parasite was reported from South Africa (De Kock et al. 1937), Chad, Congo (Malbrant et al. 1939; Pellisier et al. 1950) Sudan, Niger (Rousselot 1942) and Zimbabwe (Matson 1967).

In South Africa the parasite was discovered while examining blood slides during investigations into bovine theileriosis (De Kock et al. 1937). In Zimbabwe, *E. bovis* was detected during research into *Theileria parva lawrencei* infection (Matson 1967). In both these cases it was an incidental finding and the parasite was not considered to be pathogenic. Rousselot (1942) identified *E. bovis* in a splenectomized cow that had a persistent temperature reaction which declined immediately before the death of this animal. In 1953 Rousselot described a condition of cattle in the Sudan known as "Nofel" which was characterized by nervous signs and a high mortality. *E. bovis* was thought to be the cause of the disease. A similar disease was also reported from the Central

[1] Department of Infectious Diseases & Public Health, Faculty of Veterinary Science, Medical University of Southern Africa, 0204 Medunsa, South Africa

African Republic (Finelle 1958), Senegal (Rioche 1966) and Niger (Rousselot 1942; Girard and Rousselot 1945).

2 Clinical Aspects, Epidemiology and Treatment

2.1 Clinical Signs

Workers in Central Africa described a syndrome causing severe clinical signs and attributed this to *E. bovis* infection (Rousselot 1942; Girard and Rousselot 1945; Rousselot 1953; Finelle 1958; Rioche 1966). They recognized a peracute, acute and chronic form of the disease.

The peracute form is rare and is seen in young animals which are apparently normal in the evening and are found dead the following morning. These animals frequently bellow, froth at the mouth, the ears droop and frequent head shaking is seen. The animal starts to circle, then collapses and dies within a few hours (Rousselot 1953; Rioche 1966).

In the acute form of the disease the first clinical signs observed are the drooping of one ear, although occasionally both ears may be affected. The head is usually tilted to the affected side. Examination of the ear shows no signs of paralysis (Rousselot 1953). Head shaking is frequent; this is followed by circling and eventually the animal lies down. The animal is anorexic, the lymph nodes are enlarged and temperatures between 39.5 and 40.5 °C are recorded. The duration of the disease is usually between 36 and 48 h and death usually follows. In some cases ataxia may be seen. Rioche (1966) described depression, head pushing, bellowing, muscular tremors, opisthotonus, torticollis; death followed an epileptiform crisis. Finelle (1958) described a lacrimal discharge with congestion of the eyes and nostrils, accompanied by constipation.

Clinical signs are not apparent in the chronic form.

2.2 Pathology

Pathological changes are minimal (Rousselot 1953). These include marked hypertrophy of the lymph nodes and occasional evidence of a slight hydropericardium. In the central nervous system the medulla is strongly compressed with abundant cerebrospinal fluid, which was thought to be the cause of the nervous signs. Lesions were not detected in the ear, nor were ticks seen in the ear canal.

The drooping of the ear was thought to be due to the enlarged parotid lymph node exerting pressure on the nerve supply to the ear. Rioche (1967) described intense congestion of all serous surfaces with hemorrhagic effusions into the body cavities. Petechial hemorrhages were observed on the myocardium and oedema of the lungs was present. There was intense congestion of the meninges and of the renal cortex.

2.3 Epidemiology

The disease occurs seasonally and is associated with increased tick activity. In the Sudan, this occurred at the beginning of winter when there was a change from very hot to very cold weather (Rousselot 1953). Finelle (1958) noted that animals which had been dipped regularly did not develop the disease. Mortality can be as high as 40% in some outbreaks of the disease (Rousselot 1953).

2.4 Experimental Transmission

Donatien and Lestoquard (1936) isolated *E. bovis* by feeding *Hyalomma* on cattle. These ticks had previously fed in the larval and nymphal stages on Iranian cattle infected with *Theileria.* The incubation period was 14 to 15 days and the infection produced was of low pathogenicity. Inoculation of blood into experimental animals resulted in the development of a mild clinical disease characterized by the finding of parasites in the monocytes (Rousselot 1953). Matson (1967) was able to transmit *E. bovis* with the brown ear tick, *Rhipicephalus appendiculatus.* Parasites were detected between 20 and 41 days after attachment of adult ticks. From field observations made in South Africa, de Vos (1981) believed that *R. appendiculatus* was often infected. Norval (1979) was able to transmit *E. bovis* from field collected *R. appendiculatus.* He considered it to be mildly pathogenic. Rioche (1966) collected *Amblyomma variegatum* from the field, ground them up in Hanks balanced salt solution and injected this subcutaneously into two cows. *E. bovis* was identified between 16 and 29 days later; however, both these animals developed an inapparent form of the disease. Injection of blood into a horse yielded negative results. No parasites were detected in a 3-month-old fetus from an infected cow.

A brief parasitemia occurred in a sheep inoculated with infected blood. Two dogs infected with *E. bovis* were negative without showing a febrile reaction or parasitemia (Rousselot 1953). Donatien and Lestoquard (1940) were able to infect a monkey (*Macacus inuus*). The incubation period was 8 days followed by a temperature reaction lasting 7 days, during which parasites could be demonstrated in blood smears.

E. bovis has been isolated in South Africa from clinically normal cattle. Blood from an animal in the field was collected for isolation of *Theileria* sp. This blood was inoculated into a bovine and positive blood smears indicated that *E. bovis* infection had become established in the host. The parasite could be transmitted by intravenous inoculation of infected blood and was successfully preserved at $-170\,^{\circ}C$ using 10% DMSO as a cryoprotectant (Stewart 1976, laboratory observation[2]). Recently it has been shown that this material is still infective for cattle (Stewart and Stolsz 1988, laboratory observations[2]). In all these transmission experiments, clinical signs due to *E. bovis* were generally mild

[2] Prof C.G. Stewart, Department of Infectious Diseases and Public Health, Faculaty of Veterinary Science, Medical University of Southern Africa, 0204 MEDUNSA, South Africa

with a temperature reaction and enlargement of lymph nodes being the only consistent findings.

2.5 Morphology

The organisms multiply in the cytoplasm of monocytes. They appear as compact inclusions containing numerous organisms in a tightly packed morula. Usually only one morula is found per cell although more than one may occasionally be present. The largest morulae are 11 μm–6 μm and the smallest are 1–2 μm (Donatien and Lestoquard 1936). The smaller forms are granular in nature. This may vary from a single inculsion to many inclusions in a single monocyte (Donatien and Lestoquard 1937b). Parasites may be rare in the blood, but more numerous in smears from the lungs and liver (Donatien and Lestoquard 1937a). In blood smears from experimentally infected animals, mitotic divisions can be seen in lymphocytes at the time or just before the development of parasitemia. A monocytosis (Rousselot 1953; Rioche 1966) with eosinopaenia (Rioche 1966) is a consistent finding in infected animals and is an important factor when making a diagnosis.

2.6 Serology

Logan et al. (1986) showed that *Ehrlichia equi* and to a lesser extent *E. canis*, cross-reacted serologically with *Cowdria ruminantium*. Du Plessis et al. (1987) obtained 31 positive serological reactions to *C. ruminantium* out of 43 samples examined from one property in Namibia. As *Cowdria ruminantium* does not occur in this area, these cross-reactions were thought to be due to an *Ehrlichia* sp. In the same study only 3 out of 12 bovine sera positive to the agent of Jembrana disease and only 1 out of 12 sera from cattle that had been infected with a Kenyan strain of *Ehrlichia* gave a positive reaction to *C. ruminantium*. The *Ehrlichia* sp. from Kenya was isolated by Morzaria et al. (1985).

2.7 Diagnosis

Bovine ehrlichiosis can be recognized by the development of a temperature reaction, enlargement of the lymph nodes, ear drooping and the finding of *E. bovis* in blood or organ smears.

2.8 Treatment

Rioche (1966) successfully treated acute cases with auromycin by intravenous injection at a dose of 4 mg/kg for 4 to 5 days.

3 Discussion

Inoculation of *E. bovis* into experimental cattle has always resulted in mild clinical signs. The failure to reproduce the severe clinical disease reported from the Sudan, Senegal and the Central Africa Republic, would suggest that other factors predispose towards the development of clinical disease or that confusion occurred with other diseases. The presence of *E. bovis* in blood smears could have been an incidental finding. The main diseases which could have caused confusion are bovine cerebral theileriosis or "turning sickness" and heartwater. The clinical signs of cerebral theileriosis are similar to those described for "Nofel", except that a temperature reaction does not normally occur in cerebral theileriosis (Mettam and Carmichael 1936). In addition, the typical brain lesions seen in turning sickness were not described. In the case of heartwater, in all outbreaks of the disease a search was specifically made for *Cowdria ruminantium* with negative results.

In the Sudan "Nofel" was described as occurring at the beginning of winter when climatic conditions were unfavourable, with misty conditions and a change from very hot summer temperatures to cold winter weather. Ticks were abundant and many animals were affected by mange and sporotrichiosis. It was suggested that these conditions could have resulted in the disruption of premunition and development of the parasite, which may have accounted for the severe disease described. The fact that *E. bovis* was recorded during investigations, when tick infestations were also high (De Kock et al. 1937) may indicate that the parasitemia may increase during episodes of stress or heavy tick challenge, making it easier to demonstrate in blood slides. It is possible that premunity to *E. bovis* (Parrot 1937) is reduced under these conditions, resulting in development of parasitemia or even severe disease. These findings are similar to those of De Kock et al. (1937) who found that *Theileria mutans* infections in susceptible cattle were usually mild; however, when cattle from tick-free areas were exposed to heavy infestations by the adults of *R. appendiculatus* at the same time as infection with *T. mutans*, they developed severe infection with high mortality. This was in marked contrast to cattle which had previously been exposed to ticks and then subsequently exposed to the same challenge by *R. appendiculatus* and *T. mutans*. In the latter case no mortality occurred. The fact that *E. bovis* was also identified during these observations may be significant.

Thomas and Neitz (1958) reported similar findings with high mortality from babesiosis, anaplasmosis and *T. mutans* infection, following heavy tick challenge with *R. appendiculatus*. These deaths occurred in spite of treatment and it was suggested that the ticks were able to cause immunosuppression. It has recently been confirmed that *R. appendiculatus* is capable of suppressing host immune responses (Fivaz 1989).

The involvement of *R. appendiculatus* was, however, not described in the severe out-breaks of bovine ehrlichiosis in the Sudan. The main tick species described by Rousselot (1953) were *A. variegatum*, *R. evertsi* and *Hyalomma* sp. Finelle (1958) describes heavy tick infestation with *A. variegatum* and *Boophilus decloratus* in herds where deaths occurred due to *E. bovis* infection. He also reported that herdsmen had observed that cattle that were regularly dipped did not develop the disease.

It has recently been suggested that Jembrana disease may be similar to bovine ehrlichiosis (Ressang and Iwan 1985). In this disease reported from the island of Bali, a mortality of 10 to 20% occurs. It is associated with extensive subserosal, mucosal and endocardial hemorrhages with accumulation of yellowish to hemorrhagic fluid in body cavities. However, the clinical signs described for "Nofel" differ in some respects from those of Jembrana disease. Mucosal erosions of the mouth described in Jembrana disease were not described in the case of "Nofel" (Geering and Forman 1987). Further inoculations of the infective agent, which was grown in macrophage cultures prepared from peripheral blood, produced severe disease (Ressang and Iwan 1985). In contrast inoculation of *E. bovis* into experimental animals has always produced mild clinical signs. Rioche (1967) however described hemorrhagic effusions into the body cavities similar to Jembrana disease, however, Rousselot (1953) did not describe these lesions.

The prevalence of *E. bovis* is not known. The parasite is seldom recognised during routine examination of slides in other circumstances. When one considers the large number of slides examined annually this would suggest that under normal conditions the parasitemia is very low or that the parasite is rare.

To summarize, it can be concluded that *E. bovis* is a parasite of low pathogenicity to cattle and that severe disease only occurs when other, as yet unidentified, factors are present.

References

De Kock G, van Heerden CJ, du Toit R, Neitz WO (1937) Bovine Theileriasis in South Africa with special reference to *Theileria mutans*. Onderstepoort J Vet Sci 8:9–125

De Vos AJ (1981) *Rhipicephalus appendiculatus* cause and vector of diseases in Africa. J S Afr Vet Assoc 52:315–322

Donatien A, Lestoquard F (1936) *Rickettsia bovis*, nouvelle espéce pathogéne pour le boeuf. Bull Soc Pathol Exot 29:1057–1061

Donatien A, Lestoquard F (1937a) Particularités des *Rickettsia* des monocytes. Bull Acad Vet 10:183–187

Donatien A, Lestoquard F (1937b) Etat actuel des connaissances sur les rickettsioses animales. Arch Inst Pasteur Algér 15:142–187

Donatien A, Lestoquard F (1940) Rickettsiose bovine Algérienne a *R. bovis*. Bull Soc Pathol Exot 33:245–248

Du Plessis JL, Camus E, Oberem PT, Malan L (1987) Heartwater serology: some problems with the interpretation of results. Onderstepoort J Vet Res 54:327–329

Finelle P (1958) Note clinique rickettsiose a *Rickettsia bovis* en Oubangui-Chari. Rev Elev Méd Vet Pays Trop 11:291–292

Fivaz BH (1989) Immune suppression induced by the brown ear tick, *Rhipicephalus appendiculatus* Neumann 1901. J Parasitol 75:946–952

Geering WA, Forman AJ (1987) Animal health in Australia, vol 9. Exotic diseases. Australian Government Publishing Service, Canberra, pp 131–137

Girard H, Rousselot R (1945) La rickettsiose bovine a *Rickettsia bovis* au Soudan Francais. Bull Soc Pathol Exot 38:64–77

Logan LL, Holland DCJ, Mebus CA, Ristic M (1986) Serological relationship between *Cowdria ruminantium* and certain *Ehrlichia* species. Vet Rec 119:458–459

Malbrant R, Bayron M, Rapin P (1939) Protozooses sanguines des animaux domestiques en A.E.F. Bull Soc Pathol Exot 32:953

Matson BA (1967) Theileriosis in Rhodesia (Zimbabwe): a study of diagnostic specimens over two seasons. J S Afr Vet Med Assoc 38:93–97

Mettam RWM, Carmichael J (1936) Turning sickness, a protozoan encephalitis of cattle in Uganda. Its relationship with East Coast Fever. Parasitology 28:254–283

Morzaria SP, Irvin AD, Kocan KM, Voight WP (1985) Isolation and characterization of an *Ehrlichia* sp. from cattle in Kenya. In: Winkler H, Kakoma I, Ristic M (eds) Worksh Diseases caused by leukocytic Rickettsiae of man and animals, University of Illinois, p 30

Norval RAI (1979) Tick infestations and tick-borne diseases in Zimbabwe Rhodesia. J S Afr Vet Assoc 50:289–292

Parrot G (1937) Les rickettsioses conferent-elles l'immunite vraie ou la premunition? Arch Inst Pasteur Algér 15:188–213

Pellissier A, Troquereau P, Trinquier E (1950) Etudes sur les rickettsioses humaines et animales en Afrique Equatoriale Francaise. III Rickettsiose générale du boeuf différente de la heartwater. Bull Soc Pathol Exot 43:168–176

Ressang AA, Iwan T (1985) Jembrana disease. Its similarity to bovine ehrlichiosis. In: Winkler H, Kakoma I, Ristic M (eds) Worksh Disease caused by leukocytic Rickettsiae of man and animals. 14–16 July 1985. University of Illinois, p 39

Rioche M (1966) La rickettsiose generale bovine au Senegal. Rev Elev Méd Vét Pays Trop 19:485–494

Rioche M (1967) Lésions microscopiques de la rickettsiose générale bovine *Rickettsia* (*Ehrlichia*) *bovis* (Donatien et Lestoquard 1936). Rev Elev Méd Vét Pays Trop 20:415–427

Rousselot R (1942) Sur quelques hemocytozoone connues ou inconnues du Soudan Francais. Bull Serv Zootech Fr Epizoot Afr Occident Fr 5:210–222

Rousselot R (1953) *Rickettsia bovis*, notes de parasitologie tropicale, vol 1. Parasites du sang des animaux. Vigot Freres, Paris, pp 87–97

Thomas AD, Neitz WO (1958) Rhipicephaline tick toxicosis in cattle: its possible aggravating effect on certain diseases. J S Afr Vet Med Assoc 29:39–50

Weiss E, Moulder JW (1984) The rickettsias and chlamydias. In: Holt JG, Sharp ME, Mair NS, Sneath PHA (eds) Bergey's manual of systematic bacteriology. Williams and Wilkins, Baltimore, pp 687–739

Canine Ehrlichiosis

J. VAN HEERDEN[1]

Summary

Canine ehrlichiosis, caused by the rickettsia, *Ehrlichia canis*, is an acute, subacute or chronic tick-transmitted disease of dogs which occurs especially in subtropical and tropical parts of the world. *E. canis* has also been isolated from other canids. The often unremarkable clinical signs associated with a pathogenic infection, the many different clinical manifestations as well as concurrent occurrence with other infections, probably results in underdiagnosis of the disease.

E. canis is usually seen in the cytoplasm of circulating mononuclear cells, primarily monocytes. Different pathophysiological mechanisms are probably operational at different stages of infection. Excessive consumption and/or destruction of blood cellular components and inadequate production of these components are of importance. The diagnosis of canine ehrlichiosis should be based on history, clinical signs and the utilisation of diagnostic aids such as blood and body fluid smears, hematology, blood chemistry and serology. Treatment includes specific treatment with tetrayclines as well as supportive therapy which may include the use of intravenous polyionic fluids, a blood transfusion and hematinics. The prognosis for the treatment of acute and subacute cases is generally good. The prognosis in cases with severe pancytopaenia is very poor. Necropsy findings vary with the stage of the disease, but the most prominent findings include petechial and ecchymotic hemorrhages on visceral surfaces of internal organs, and subcutaneous hemorrhages. Prevention of the disease can be achieved by effective tick control and by prophylactic use of tetracyclines.

1 Introduction

Canine ehrlichiosis is an acute, subacute or chronic tick-borne rickettsial disease of dogs which occurs particularly in subtropical and tropical areas of the world. The disease, caused by *Ehrlichia canis* has been described under various names: canine typhus; rickettsiosis; canine hemorrhagic fever; idiopathic hemorrhagic syndrome, tropical canine pancytopenia, tracker dog disease, Lahore canine fever and Nairobi bleeding disease (Greene and Harvey 1984).

Since the first reports of canine ehrlichiosis in Algeria in 1935 and 1936 (Donatien and Lestoquard 1935, 1936), the disease has been described in many parts of the world (Huxsoll 1976; Greene and Harvey 1984). Although, primarily of importance in countries with a relatively warm climate, the infection is also encountered in areas with a more temperate to cold climate (Keefe et al. 1982). In southern Africa the disease has been described in Namibia, Zimbabwe and South Africa. In South Africa it is especially common in the higher rainfall

[1] Department of Companion Animal Medicine and Surgery, Faculty of Veterinary Science, Medical University of Southern Africa, Medunsa 0204, South Africa

northern and eastern areas where ticks, in general, are more abundant. It also occurs commonly in dogs in the poorer developing communities in the more subtropical areas of southern Africa. Tick infestations are common on these dogs, which are usually allowed to roam freely or which may be confined to relatively small areas, where hygiene is often poor. Problems associated with these communities, such as ignorance, poverty and unavailability of veterinary health care contribute to the high incidence of tick-borne diseases (van Heerden 1989).

The often unremarkable clinical signs associated with a pathogenic infection, as well as the fact that it frequently occurs concurrently with other infections, probably result in the underdiagnosis of this disease.

2 Taxonomy, Epidemiology and Pathogenesis

2.1 The Organism

Ehrlichia canis is the type species for a group of small pleomorphic organisms which are found in the white blood cells of their respective mammalian hosts (Huxsoll 1976). *E. canis* or *Rickettsia canis* which was first described by Donatien and Lestoquard (1936) from dogs in Algeria, is taxonomically described as follows (Ewing 1969):

Division: Protophyta Sachs, 1874, emend. Krassilnikov, 1949.
Class: Microtatobiotes Philip, 1956.
Order: Rickettsiales Buchanan and Buchanan, 1938, emend. Gieszczkiewicz, 1939.
Family: Rickettsiaceae, Pinkerton, 1936
Tribe: Ehrlichieae, Philip, 1957
Genus: *Ehrlichia*, Moshkovskii, (1937), 1945
Species: *Ehrlichia canis* (Donatien and Lestoquard 1935)

E. canis is usually seen in the cytoplasm of circulating mononuclear cells, primarily in monocytes. Three different forms of the parasite, namely morulae, amorphous inclusions and small pleomorphic azurophilic inclusions have been demonstrated in buffy coat and tissue smears by Carter et al. (1971). Ultrastructural studies of these cytoplasmic inclusions have shown them to be membrane-lined vacuoles containing elementary bodies varying in size and number. Elementary bodies are surrounded by trilamellar membranes (Hildebrandt et al. 1973a; Simpson 1974). Elementary bodies (individual ehrlichia organisms), initial bodies (immature organismal inclusions) and morulae (mature organismal inclusions) have been described as different developmental forms of *E. canis* in cultured dog monocyte cells (Nyindo et al. 1971).

No conclusive information on the properties of different strains of *E. canis* is available. A reportedly mildly pathogenic strain which produced inclusions in neutrophils and eosinophils (Ewing et al. 1971) could have been *E. equi* (Lewis et al. 1975) which produced mild disease with inclusions in granulocytic cells in dogs. Leeflang and Perie (1972) could not find any difference in pathogenicity in dogs between strains of *E. canis* from different parts of the world.

Active multiplication of *E. canis* takes place in the tick. Different morphological forms of *E. canis*, similar to those described in canine monocyte cell cultures, have been detected in the midgut, salivary glands and hemocytes of partially engorged adult ticks infected as larvae and nymphs (Smith et al. 1976). Intracytoplasmic inclusions were found to contain between 1 to 80 elementary bodies. Distinct membranes surrounded each elementary body.

2.2 Transmission

The brown dog tick *Rhipicephalus sanguineus* has a worldwide distribution between latitudes 50 °N and 35 °S and is the only known vector of *E. canis*. Transmission occurs transstadially only. Transovarial transmission has not been confirmed (Groves et al. 1975).

Smith et al. (1976) found that prior partial feeding of nymphs, infected as larvae with *E. canis*, was highly desirable when ground-up ticks were used to experimentally transmit the parasite to dogs. Saliva, which is injected into the host throughout feeding by adult ticks, promotes the accumulation of monocytes and other leukocytes at the site of the tick bite and is probably the major mechanism involved in the transfer of *E. canis* from ticks to dogs. Adult *R. sanguineus* ticks have been found to transmit effectively *E. canis* for up to 155 days after having fed on a dog in the acute stage of ehrlichiosis (Lewis et al. 1977).

Experimental infection of dogs has been achieved by intravenous or intraperitoneal injection (Huxsoll et al. 1972) of infected blood, the parenteral injection of emulsified infected ticks, and by allowing infected ticks to feed on susceptible dogs. Experimental infection has also been achieved with subinoculation of blood from chronically infected animals but these animals appear to play a less important role as natural reservoirs of the disease (Lewis et al. 1977). Ticks primarily become infected within the first two weeks of feeding on dogs in the acute stage of the disease. *R. sanguineus* can survive up to 568 days without feeding and may thus act as an important resevoir of infection.

2.3 Host Range

Canine ehrlichiosis is essentially a disease of the domestic dog. Although a large number of breeds and cross-bred dogs have been found to be susceptible, German Shepherd dogs appear to be particularly affected. It appears that the disease is more likely to progress to a chronic fatal infection in this breed (Klopfer and Nobel 1972; Huxsoll 1976; Stephenson and Ristic 1978; Nyindo et al. 1980; van Heerden 1982).

Apart from the domestic dog, *E. canis* has also been isolated from other members of the family Canidae. Neitz and Thomas (1938) successfully transmitted the parasite to the blackbacked jackal *Canis mesomelas*. The jackal, however, remained asymptomatic. The parasite was subsequently demonstrated in blood and lung-tissue smears from another captive blackbacked jackal which was heavily infested with ticks and which also suffered from a concomitant *Babesia canis* infection (Neitz and Steyn 1947). Van Heerden (1979) also successfully infected blackbacked jackals which remained asymptomatic. Mild decreases in hemoglobin concentration and red cell count were observed in these animals. Price and Karstad (1980) demonstrated the presence of *E. canis* in free-ranging blackbacked jackals in Kenya by using an in vitro culturing technique. The blackbacked jackal, which commonly occurs in southern Africa, may thus play an important role in the epidemiology of the disease.

Ewing et al. (1964) successfully infected a coyote *Canis latrans frustrar* with *B. canis* and *E. canis*. Successful subinoculation of a domestic dog puppy with blood from the coyotes was achieved. Similarly, successful infection of the red fox *Vulpes fulva* and gray fox *Urocyon cinereoargenteus* was achieved. Despite the development of a mild anemia, thrombocytoenia and leukopenia in the acute stages of the disease, the foxes remained asymptomatic (Amyx and Huxsoll 1973).

A diagnosis of canine ehrlichiosis in captive wolves and wolf-dog crosses was based on clinical signs, hematological and autopsy findings (Harvey et al. 1979).

Stevenson-Hamilton (1939) suspected ehrlichiosis to be responsible for mortality among free-living wild dogs *Lycaon pictus* in the Kruger National Park. Neitz and Thomas (1938) confirmed the presence of *E. canis* in domestic dogs in the Kruger National Park but were unable to confirm the presence of the disease in wild dogs. De Kock (1935), Pienaar (1963) and Van der Merwe (1959) stated, without adequate proof, that wild dogs are susceptible to *E. canis*. Van Heerden (1979), however, confirmed the susceptibility of this species to ehrlichiosis by successfully infecting captive animals with *E. canis*. Infected wild dogs suffered a loss in body condition, listlessness, partial to complete anorexia and diarrhoea. Morulae were readily demonstrated in the cytoplasm of circulating monocytes. Laboratory investigations revealed a mild anemia, leukopenia and thrombocytopenia and all dogs recovered uneventfully after treatment with tetracyclines.

Donatien and Lestoquard (1935, 1936) claimed to have successfully infected the primate *Macacus inuus*. However, Groves et al. (1975) were unable to infect monkeys and these observations were supported by van Heerden and Goosen (1981) who were unable to experimentally infect captive vervet monkeys *Cercopithecus pygerythrus*.

Occasional tick bites in humans have been associated with symptoms of rigor, headache, nausea and anorexia as well as leukopenia, thrombocytopenia and a rise in antibody titre to *E. canis*. Patients were successfully treated with tetracyclines (Fishbein et al. 1987; Maeda et al. 1987; Caruana 1988).

2.4 Pathogenesis of Disease

Both the acute and chronic stages of the disease may be characterized by thrombocytopenia, leukopenia and anemia. Different pathophysiological events are probably of primary importance at the different stages of infection of a dog. Despite numerous investigations, our knowledge is still incomplete and it appears that excessive consumption and/or destruction of blood cellular components and inadequate production of these components, respectively, are of primary importance in the acute and chronic stages of the disease.

The organism is taken up by mononuclear cells following its transmission by ticks (Greene and Harvey 1984). *E. canis* multiplies within these cells and is transported by them to all mononuclear-phagocytic tissues in the body. Infected cells may thus become trapped in the micro-circulation of the lungs, kidneys and meninges for example where they may become attached to or invade capilliary endothelial cells. This, as well as the migration of infected cells into subendothelial tissues, may induce inflammatory reactions. These reactions together with immune-mediated reactions may be involved in consumption of blood cellular elements. Increased platelet destruction (Smith et al. 1975; Pierce et al. 1977), complement activation (Lovering et al. 1980) and enhanced phagocytosis with hyperplasia of mononuclear systems may all be involved in the development of thrombocytopenia in acute ehrlichiosis (Smith et al. 1975; Reardon and Pierce 1981).

Acute infections with *E. canis* cause hyperplasia of both B and T cell compartments of the lymphoreticular system. An optimal orchestrated response of both B and T cells is probably essential to ensure maximal immunity to *E. canis*. Kakoma et al. (1980) proposed antibody independent (direct) and antibody-dependent (indirect) immunologic pathways in canine ehrlichiosis. The high susceptibility of German Shepherd dogs may be due to an inherent breed related inability of blast formation and in the production of leukocyte migration inhibition factor (direct pathways). Cytodestructive mechanisms probably contribute to the elimination of parasitized white blood cells from the peripheral circulation. It has been shown that during the acute phase of the disease, autologous lymphocytes show lymphocytotoxic activity against host monocytes (Kakoma et al. 1977).

The same authors also observed an antibody dependent lymphocyte-mediated cytotoxity against platelets (Kakoma et al. 1980). Indirect lymphocyte participation in canine ehrlichiosis involves production of specific IgG and platelet migration inhibition factor which may accelerate platelet aggregation and sequestration.

Increased myeloid to erythroid ratio's and numbers of megakaryocytes were indicated by differential counts of bone marrow smears of dogs in the acute stage of the disease. The total cellularity of the bone marrow was not impaired (Buhles et al. 1975).

The subacute or subclinical phase of the disease is characterized by mild symptoms and the persistence of antigenic stimulation of effector cells in the body. Immunocompetent animals may well be able to halt the further development of disease processes.

The chronic stage of the disease developes in immunologically compromised animals. Depressed cell-mediated immunity has been proposed as a possible reason for the development of the chronic form of the disease in German Shepherd dogs (Nyindo et al. 1980).

The progressive increase in production of immunoglobulins A, M and G during the course of the disease results in high serum concentrations of gamma globulins in the chronic stage (Weisiger et al. 1975). These antibody titres do not seem to be indicative of resistance to the infection and may in fact contribute in an as yet unknown way to the development of disease in the patient.

The chronic stage of the disease is further characterized by the development of bone marrow hypoplasia which results in anemia, leukopenia and thrombocytopenia. The cause of the bone marrow hypoplasia is unknown. Stem cell infection or an immunologic injury to the microcirculation have been proposed as possible mechanisms (Buhles et al. 1975). Persistence or progression of the bone marrow hypoplasia is apparently not dependent on persistence of infection with *E. canis*. The severe plasma cell infiltration into organs such as the kidneys, spleen and meninges may further injure the health of chronically infected dogs by resulting in general malaise, inappetence and a proteinuria.

3 Clinical Signs of Disease

Various authors have suggested that purebred dogs appear to be more susceptible to *E. canis* than mixed breed dogs (Raghavachari and Reddy 1958; Seamer and Snape 1972; Stephenson and Ristic 1978; van Heerden, 1982). German Shepherd dogs, in particular, have been found to be relatively more susceptible than Beagle dogs (Nyindo et al. 1980; Huxsoll et al. 1972; Klopfer and Nobel 1972; Seamer and Snape 1972) and to be immunologically less responsive than Beagle dogs. Other investigators however, failed to detect any breed tendencies (Kuehn and Gaunt 1985).

Various forms and/or stages of the disease have been described. Carmichael and Fiennes (1942) recognized three forms of canine typhus (canine rickettsiosis or canine ehrlichiosis): a cutaneous form, a septicaemic form and a nervous form. Confusion by these authors of canine ehrlichiosis with other diseases such as canine distemper is unfortunately not beyond doubt.

Walker et al. (1970) describe febrile, subclinical and terminal phases of the disease. Dogs in the terminal phase which followed on the subclinical phase were classified as dogs with pancytopaenia and those with epistaxis. The latter phase was again subdivided into acute and chronic categories.

Buhles et al. (1974) referred to mild chronic (persistence of infection without clinical signs) and severe chronic tropical canine pancytopenia. Van Heerden (1982) proposed the terms acute, subacute and chronic to describe the different stages of the disease. Subclinical implies the complete absence of clinical signs which does not apply strictly to the stage that follows on the acute stage of the disease. Clinical signs in this stage are often minimal and subtle and may only be revealed on close inspection and prolonged observation of the patient. The term "subclinical ehrlichiosis" should be reserved to describe a state of antigenic persistence (infection with *E. canis*) without clinical signs. Codner and Farris-Smith (1986) found a high serum globulin concentration to be a non-specific sensitive indicator of subclinical infection with *E. canis*. Such a state may relapse into clinical ehrlichiosis probably depending on factors such as nutritional status, immunocompetence and concurrent infections.

3.1 Acute Stage

Clinical signs of disease are seen after a prepatent period of 4–21 days (Bool and Sutmoller 1957; Van Dijk 1971; van Heerden and van Heerden 1981). Considerable variation occurs in both the expression and duration of clinical signs. The acute stage of disease is often not apparent. The most common clinical presentation is partial anorexia, malaise, an elevated rectal temperature, a mildly elevated pulse rate and congested mucous membranes. Serous to purulent oculonasal discharges (Seamer and Snape 1972), dyspnea, increased lung sounds, enlargement of peripheral lymph nodes (van Heerden and van Heerden 1981), splenomegaly and nervous signs such as ataxia, hindquarter paresis and cranial nerve deficits may also be seen. Lameness due to a polyarthritis caused by *E. canis* have been described by Cowell et al. (1988). Edema of the legs and scrotum, vomition, corneal opacity, conjunctivitis, hyphema and mild epistaxis have been observed in dogs in what has been described by Walker et al. (1970) as the febrile phase of the disease. Cardiac dysrhythms, such as ventricular tachycardia, have been encountered occasionally by the author in infected dogs.

3.2 Subacute Phase

The most common clinical signs in the subacute phase of the disease are a loss in body condition, intermittent malaise and anorexia, a mildly elevated pulse rate, enlargement of peripheral lymph nodes and pale mucous membranes. Petechial hemorrhages may be found in the skin. In areas where the stable fly *Stomoxys calcitrans* occurs, dogs are often presented with dry clotted blood on the ears. The dogs may also be infected with large numbers of ticks.

3.3 Chronic Phase

In the chronic stages of the disease, most dogs would typically present with overt signs of a marked loss in body mass. The dog often has a history of intermittent epistaxis and anorexia. Some patients retain their body condition until the terminal stages of the disease. Pale mucous membranes in association with a relatively weak pulse and an increased heart rate are commonly found. Dry clotted blood is seen on the ear pinnae, petechial hemorrhages may be seen in the skin and mucous membranes and the dog may show epistaxis. Ulceration of the skin with secondary infections occur especially over bony prominences.

Non-inflammatory edema of the distal parts of the legs and scrotum, dyspnea, fever, ocular lesions, lymphadenopathy, posterior weakness and splenomegaly are sometimes seen (Huxsoll et al. 1970a, b). Meinkoth et al. (1989) reported multiple episodes of grand mal seizures in a dog with clinical signs and laboratory findings suggestive of canine ehrlichiosis.

Ocular lesions may include tortuous retinal vessels, dark gray spots in the tapetal area (Ellett et al. 1973), subretinal hemorrhage, retinal detachment anterior uveitis, hyphema and corneal opacities (Troy et al. 1980).

Kidney disease which may result in uremia and which is characterized by polyuria, polydypsia, vomition and ulcerations of the buccal mucosae may be seen in the chronic stages of the disease.

Infected bitches may bleed excessively during pro-estrus and during the post-partum period. Abortion, inability to conceive and neonatal deaths have also been observed by the author.

3.4 Concurrent Disease

Concurrent infection of *E. canis* infected dogs with *Babesia canis*, *Hepatozoon canis*, *Leishmania donovani* and *Cryptococcus neoformans*, have been described (Neitz and Thomas 1938; Mudaliar 1944; Immelman and Button 1973; van Heerden 1979, 1982; Troy et al. 1980; Adeyanju and Aliu 1982; Ezeokoli et al. 1983; Gossett et al. 1984; Schaer et al. 1985; Collett et al. 1987; Price et al. 1987).

Experimental infections of dogs with combined *E. canis* and *B. canis* resulted in severe disease, while experimental infection of dogs with a preimmunity to *B. canis*, with *E. canis* resulted in an increase in parasitaemia (Ewing and Buckner 1965). Mixed infections with *E. canis*, *B. canis*, *H. canis*, and/or other pathogens such as canine distemper virus, canine parvo virus and the hookworm *Ancylostoma caninum*, are commonly encountered by the author. Clinical signs in these mixed infections vary widely but anaemia and the presence of dry clotted blood on the surface of the ears are commonly seen. Pin-pricking of the ears in these patients often results in excessive bleeding.

Mixed infections of *B. canis* and *E. canis* may easily be overlooked (van Heerden and Immelman 1979). The relative ease with which the former parasite is identified on a blood smear may well result in overlooking the presence of *E. canis* which is often only demonstrated after careful and prolonged examination. So-called biliary fever-relapse cases should be investigated in particular for the presence of an underlying *E. canis* infection.

Although *Hepatozoon canis* is normally regarded as non-pathogenic, clinical signs of disease such as anorexia, general weakness, lethargy, pale mucous membranes, fever and epistaxis have been ascribed to infections with this parasite (Ezeokoli et al. 1983). Severe chronic ehrlichiosis has also been associated with severe parasitemia with *H. canis* (van Heerden 1981).

Evidence of a loss in body mass, enlarged peripheral lymph nodes, pale mucous membranes, epistaxis and osteoarthritic crepitation in both coxofemoral joints were found in a dog infected with both *Leishmania donovani* and *E. canis* (Schaer et al. 1985). General weakness, loss in body mass, poor appetite and pale mucous membranes were observed in a dog infected with both *Cryptococcus neoformans* and *E. canis* (Collett et al. 1987)

4 Diagnosis

The diagnosis of canine ehrlichiosis should be based on history, clinical signs and the utilization of diagnostic aids such as the examination of blood smears, body fluids, hematology, blood chemistry and serology.

4.1 History

The area of origin of the patient, infestation with *Rhipicephalus sanguineus* and a history of intermittent anorexia, loss in body condition, intermittent malaise and epistaxis could be indicative of canine ehrlichiosis.

4.2 Clinical Signs

The clinical signs in canine ehrlichiosis are generally non-specific. Any dog with a fever reaction, pale mucous membranes, epistaxis or with evidence of chronic progressive weight loss in an area where the disease is endemic, should be investigated for the possibility of an infection with *E. canis*.

4.3 Microscopic Examination of Blood Smears

Careful and systematic examination of stained peripheral blood smears may reveal the presence of *E. canis* morulae or granules in the cytoplasm of monocytes. Morulae have been seen from 11 to 20 days after infection in a group of experimental dogs, 2–11 days after the first rise in rectal temperature (van Heerden and van Heerden 1981). Morulae have, in general, been found in experimental dogs more readily during the febrile reaction (van Heerden el al. 1983). It may be very difficult or impossible to demonstrate morulae in blood smears of dogs in the subclinical and chronic stages of the disease. Examination of buffy coat smears may be useful.

Examination of blood smears may reveal a relative abundance of monocytes which are large and irregular in shape with prominent vesicular nuclei and a foamy vacuolated cytoplasm. There may be evidence of reduced numbers of platelets, mild red cell regeneration (polychromasia, anisocytosis, reticulocytosis, normoblasts) in the acute stage but scant evidence of regeneration of red blood cells in the chronic stage of the disease (Buckner and Ewing 1967). Increase in size of platelets (Smith et al. 1975) or platelets with densely stained centers and clear ragged peripheries may be seen (Seamer and Snape 1972).

When examining blood smears, *E. canis* should be differentiated from phagocytozed *Babesia* parasites and bacteria, clusters of blood platelets, *H. canis*, *L. donovani*, stain deposits or artefacts superimposed on white blood cells and nuclear material (Neitz and Thomas 1938). The parasite may also be demonstrated in smears of lung, lymph node and spleen aspirates of infected animals (Donatien and Lestoquard 1936).

Price and Dolan (1980) described a technique wherein leukocyte-enriched plasma from a patient is incubated for up to 96 h to facilitate demonstration of the organism in stained preparations.

4.4 Hematology

A reduction in hematocrit, hemoglobin concentration and red cell count may be seen in all stages of ehrlichiosis. In the acute stage of the disease, an increase in reticulocyte count may be observed (Buckner and Ewing 1967) while poor reticulocyte response is encountered in the chronic stage of the disease (Walker et al. 1970). The acute and chronic stages of the disease are further characterized by a leukopenia and a thrombocytopenia (Pierce et al. 1977; van Heerden 1982), although the leukopenia, in the acute stage of the disease, is relatively mild. The total white cell count returns to normal during the subacute phase. A variable thrombocytopenia is seen in all stages of ehrlichiosis. Monocytosis may occur in the acute and chronic stages (Hibler and Greene 1986). Lymphopenia and eosinopenia were commonly found in a series of cases investigated by Kuehn and Gaunt (1985). Increases of up to 50 mm in the erythrocyte sedimentation rate have been recorded in the acute stage of infection (Seamer and Snape 1972). A total white blood cell count of less than $2000 \times 10^9 \, l^{-1}$ is regarded as evidence of a very poor prognosis (van Heerden 1982).

The Coombs test has been found positive in a few dogs (Troy et al. 1980; Kuehn and Gaunt 1985).

Bone marrow examination during the acute and chronic stages of the disease reflects a hypercellularity of the megakaryocytic and myeloid series, and hypocellularity of the megakaryocytic, myeloid and erythroid series respectively. Marrow plasmacytosis occurs at all stages of the disease (Hibler and Greene 1986).

4.5 Blood Chemistry

Hypergammaglobulinemia is a constant finding in subacute and chronic cases of ehrlichiosis (Buhles et al. 1974). Burghen et al. (1971) described a hypergammaglobulinemia, hypergammaglycoglobulinemia and a decrease in serum albumin concentration. The acute stage of the disease was characterized by slight increases in alpha 2 proteins and glycoglobulins.

Electrophoretic examination of serum of infected dogs may occasionally reveal a monoclonal hypergammaglobulinemia (Hibler and Greene 1986; Matus et al. 1987).

Elevated serum urea (Walker et al. 1970), creatinine, and phosphorus concentrations may be encountered in chronic cases of ehrlichiosis (Kuehn and Gaunt 1985) while the serum concentration of alkaline phosphatase and alanine transaminase may occasionally be elevated (Walker et al. 1970; Kuehn and Gaunt 1985). Proteinuria is often found in dogs with chronic ehrlichiosis (Kuehn and Gaunt 1985; Troy et al. 1980)

4.6 Analysis of Cerebrospinal and Joint Fluids

Mild increased protein concentration in the cerebrospinal fluid has been reported. Large mononuclear cells of which some contained intracytoplasm morulae of *E. canis* were found in a dog with clinical and laboratory findings typical of canine ehrlichiosis (Meinkoth et al. 1989). *Ehrlichia* morulae were seen in neutrophils in joint fluid from dogs with polyarthritis (Cowell et al. 1988).

4.7 Serological Diagnosis

The indirect fluorescent antibody technique which uses *E. canis*-infected cell culture as the antigen has been found to be highly sensitive and specific for *E. canis* (Ristic et al. 1972). Antibodies to *E. canis* may be detected as early as 7 days after the initial infection (Weisiger et al. 1975). The same authors demonstrated that the earliest antibodies to *E. canis* in experimentally infected dogs were primarily immunoglobulins M and A. The immunoglobulin G proteins then gradually increased. A drop in antibody titre was experienced in some dogs prior to death.

E. canis has also been demonstrated in circulating white blood cells by direct fluorescence of buffy coat smears with specific fluorescent antibody (Carter et al. 1971).

5 Treatment

5.1 Specific Treatment

Early investigators reported on the use of chloramphenicol, procaine penicillin G, sulfadimethoxine, sulfacetamide, sulfapyridine, formalinized saline, oxytetracycline and tetracycline in clinical cases of canine ehrlichiosis (Buckner and Ewing 1967; Ewing 1969). Controlled studies to evaluate the efficacy of these drugs were not conducted. Oral administration of oxytetracycline at a dosage of 66 mg kg^{-1} for 14 days resulted in remission of clinical signs and in the sterilization of the infection in 13 out of 15 dogs (Amyx et al. 1971). Dogs in which the infection was sterilized, did not develop resistance to *E. canis* and could easily be re-infected (Amyx et al. 1971; Buhles et al. 1974). The drug was also found to be an effective prophylactic agent when administered orally at a dosage rate of 6.6 mg kg^{-1} day^{-1} (Amyx et al. 1971). Relapses may occur with the use of oxytetracycline parenterally at a dosage rate of 10 mg kg^{-1} for 10 days (Immelman and Button 1973). Troy et al. (1980) used tetracycline hydrochloride orally at 17.5 mg kg^{-1} three times daily. Oxytetracycline has also been used effectively at dosage rates of 100 mg kg^{-1} orally, once a day or at 50 mg kg^{-1} twice daily for not less than 10 days (van Heerden 1982). The intramuscular use of oxytetracycline usually results in very painful swellings and is not recommended. The use of doxycycline orally or intravenously at dosages ranging from 4.6 to 16.7 mg kg^{-1} resulted in remission of clinical signs in clinical cases (van Heerden and Immelman 1979). Dosages of 2–10 mg kg^{-1} for 10 days resulted in remission of clinical signs in experimental cases, but relapses occured. Sterilization of infection occurred in some dogs at dosages ranging from 4–10 mg kg^{-1} (Williams, van Heerden and Evezard 1989, unpubl.).

It is the author's belief that dogs should be treated, either orally or intravenously for not less than 10 days, with doxycycline at a dosage rate of 10 mg kg^{-1}. Cases which relapse should be treated for a longer period of time with low dosages of either doxycycline or oxytetracycline after the initial course of treatment.

Price and Dolan (1980) reported successful treatment of clinical cases with imidocarb diproprionate at a dosage rate of 5–7 mg kg^{-1} intramuscularly, twice with an interval of 14 days. Dogs were tested for persistence of infection after treatment by using a cell-culture test. Ogunkoya et al. (1981) found imidocarb to be less effective against *E. canis* than against *B. canis*, *H. canis* or mixed infections of these parasites. Adeyanju and Aliu (1982) reported a good clinical response after treatment with a single intramuscular dosage of imidocarb diproprionate at 5 mg kg^{-1}. Van Heerden and Van Heerden (1981) were, however, unsuccessful in sterilizing the infection with imidocarb diproprionate in experimentally infected dogs at a dosage rate of 6 mg kg^{-1} administered subcutaneously and repeated after 14 days. Subinoculation of blood from treated dogs to susceptible animals resulted in overt signs of ehrlichiosis. The author has also found the drug to be ineffective in clinical cases of ehrlichiosis. The use of the drug seems to be controversial (Villemin et al. 1984).

5.2 Supportive Treatment

Supportive therapy may include the use of intravenous polyionic fluids, a blood transfusion and hematinics in case of severe and chronic blood loss. Thrombocytopenic anemic dogs in our clinic are routinely treated with a fresh blood transfusion. Blood is used immediately after collection because of a lack of facilities to separate and store platelets or platelet-rich plasma.

The prognosis for severe chronic cases of ehrlichiosis with a total white cell count of less than $2 \times 10^9\ l^{-1}$ is poor. Corticosteroids, although recommended for use early in the treatment of ehrlichiosis by some authors, is in my opinion, contraindicated as they result in exacerbation of the condition of the patient. Long-term or excessive use of corticosteroids are contraindicated in any event because dogs are usually leukopenic and immunosuppressed (Hibler and Greene 1986). The intravenous administration of vincristine (weekly at a dosage of 0.010 to 0.025 mg kg^{-1}) or vincristine-loaded platelets may be considered in the treatment of bleeding patients. Bone marrow stimulation with anabolic steroids may also be considered (Hibler and Greene 1986). Successful long-term treatment of two severely pancytopenic dogs included the use of levamisole at daily oral dosages between 3–10 mg kg^{-1} for 60 days (van Heerden 1982).

Monoclonal gammopathy associated with *E. canis* infection was successfully treated with cyclophosphamide (7 mg kg^{-1}, intravenous) and melphalan

(0.1 mg kg^{-1}, 24 h, per os) and prednisolone to suppress the immunoglobulin response. An estimated volume of plasma (body mass in kg × (100-PCV) × 0.08) was removed and replaced with Ringers lactate. Melphalan and prednisolone treatment was continued for 30 days after plasmaphoresis (Matus et al. 1987).

Treatment for concomitant infections should include the specific chemotherapeutic agents. However, carriers of *B. canis* which were experimentally infected with *E. canis* and which subsequently developed clinical signs and hematological evidence of ehrlichiosis, fully recovered despite treatment with doxycycline only (van Heerden et al. 1983). The decision not to treat the *B. canis* infection in mixed *B. canis/E. canis* infections in clinical cases is a difficult one. This should, however, be considered in so-called biliary-relapse cases which are subsequently shown to be infected with both *E. canis* and *B. canis.*

The prognosis for the treatment of acute and subacute ehrlichiosis is generally good. The prognosis for chronic cases with severe pancytopenia is very poor and the prognosis is regarded as almost hopeless once the total white cell count has dropped to below $2 \times 10^9 l^{-1}$, especially if the dog is looking weak and depressed.

6 Post-Mortem Findings

Necropsy findings will vary with the stage of the disease. The most prominent findings include petechial and ecchymotic hemorrhages on visceral surfaces of internal organs as well as subcutaneous hemorrhages. In chronic cases, evidence of emaciation may be seen. The bone marrow appears red in the acute and pale in the chronic stage of infection.

Lymphoreticular hyperplasia in the paracortical area of lymph nodes, interstitial pneumonia, subendothelial aggregates of mononuclear cells in pulmonary blood vessels, renal periglomerular and perivenular plasmacytosis, hemopoetic hyperplasia and perivascular cuffing by lymphocytes and plasma cells in many organs were lesions typical of the acute stage of infection in experimental dogs (Van Dijk 1971; Reardon and Pierce 1981).

Prominent histological features in chronic ehrlichiosis are plasma cell infiltration of meninges, kidneys and lymph nodes (Hildebrandt et al. 1973b), bone marrow atrophy (Huxsoll et al. 1970a, b; Buhles et al. 1975), and glomerulonephritis (Troy et al. 1980).

7 Prevention

Prevention of the disease can be achieved by effective tick control and by the prophylactic use of a tetracycline. Tick control should include regular dipping of dogs and spraying of kennels with an effective acaricidal preparation. Absolute tick control may be impossible in dogs in areas where ticks are common, especially in dogs working in rural areas. The prophylactic use of tetracycline under these circumstances at oral dosages of 6.6 mg kg^{-1} or a total dose of 250 mg daily for prolonged periods of time has been reported (Willder 1977;

Davidson et al. 1978). Such long-term treatment with tetracycline has not been found to result in impairment in conception, gestation, parturition, lactation, litter size or litter health in a unit of approximately 60 dogs over a period of 4 years (Willder 1977). Clinical cases which continuously relapsed after treatment with a tetracycline have also been treated prophylactically for prolonged periods with low dosages of either doxycycline or tetracycline with favorable results. (van Heerden, unpubl data).

References

Adeyanju BJ, Aliu YO (1982) Chemotherapy of canine ehrlichiosis and babesiosis with imidocarb. J Am Anim Hosp Assoc 18:827–830

Amyx HL, Huxsoll DL (1973) Red and gray foxes – potential reservoir hosts for *Ehrlichia canis*. J Wildl Dis 9:47–50

Amyx HL, Huxoll DL, Zeiler DC, Hildebrandt PK (1971) Therapeutic and prophylactic value of tetracycline in dogs infected with the agent of tropical canine pancytopaenia. J Am Vet Med Assoc 159:1428–1432

Bool PH, Sutmoller P (1957) *Ehrlichia canis* infections in dogs on Aruba (Netherlands Antilles). J Am Vet Med Assoc 130:418–420

Buckner RG, Ewing SA (1967) Experimental treatment of canine ehrlichiosis and haemobartonellosis. J Am Vet Med Assoc 150:1524–1530

Buhles WC, Huxsoll DL, Ristic M (1974) Tropical canine pancytopaenia: Clinical, hematologic, and serologic response of dogs to *Ehrlichia canis* infection, tetracycline therapy, and challenge inoculation. J Infect Dis 130:357–367

Buhles WC, Huxsoll DL, Hildebrandt PK (1975) Tropical canine pancytopaenia: role of aplastic anaemia in the pathogenesis of severe disease. J Comp Pathol 85:511–521

Burghen GA, Beisel WR, Walker JS, Nims RM, Huxsoll DL, Hildebrandt PK (1971) Development of hypergammaglobulinaemia in tropical canine pancytopaenia. Am J Vet Res 32:749–756

Carmichael J, Fiennes RNTW (1942) Rickettsia infection of dogs. Vet Rec 54:3–4

Carter GB, Seamer J, Snape T (1971) Diagnosis of tropical canine pancytopaenia (*Ehrlichia canis* infection) by immunofluorescence. Res Vet Sci 12:318–322

Caruana LB (1988) *Ehrlichia canis* infections in humans. Clin Lab Sci 1:283–284

Codner EC, Farris-Smith LL (1986) Characterization of the subclinical phase of ehrlichiosis in dogs. J Am Vet Med Assoc 189:47–50

Collett MG, Doyle AS, Reyers F, Kruse T, Fabian B (1987) Fatal disseminated cryptococcosis and concurrent ehrlichiosis in a dog. J S Afr Vet Assoc 58:197–202

Cowell RL, Tyler RD, Clinkenbeard KD, Meinkoth JH (1988) Ehrlichiosis and polyarthritis in three dogs. JAVMA 192:1093–1095

Davidson DE, Dill GS, Tingpalapong M, Premabutra S, Nguen PL, Stephenson EH, Ristic M (1978) Prophylactic and therapeutic use of tetracycline during an epizootic of ehrlichiosis among military dogs. J Am Vet Med Assoc 172:697–700

De Kock G (1935) Wild animals as carriers of infection. S Afr J Med October:725–730

Donatien A, Lestoquard F (1935) Existence en Algérie d'une *Rickettsia* du chien. Bull Soc Pathol Exot 28:418–419

Donatien A, Lestoquard F (1936) Existence de la premunition dans le rickettsiose naturelle ou experimentale du chien. Bull Soc Pathol Exot 29:378–383

Ellett EW, Player RF, Pierce KR (1973) Retinal lesions associated with induced canine ehrlichiosis: a preliminary report. J Am Anim Hosp Assoc 9:214–218

Ewing SA (1969) Canine ehrlichiosis. Adv Vet Sci Comp Med 13:331–353

Ewing SA, Buckner RG (1965) Manifestations of babesiosis, ehrlichiosis and combined infections in the dog. Am J Vet Res 26:815–828

Ewing SA, Buckner RG, Stringer BG (1964) The coyote, a potential host for *Babesia canis* and *Ehrlichia* sp. J Parasitol 50:704

Ewing SA, Roberson WR, Buckner RG, Hayat CS (1971) A new strain of *Ehrlichia canis* J Am Vet Med Assoc 159:1771–1774
Ezeokoli CD, Ogunkoya AB, Abdullahi R, Tekdek LB, Sannusi A, Ilemobade AA (1983) Clinical and epidemiological studies on canine hepatozoonosis in Zaria, Nigeria. J Small Anim Pract 24:455–460
Fishbein DB, Sawyer LA, Holland CJ, Hayes EB, Okoroanyanwu W, Williams D, Sikes K, Ristic M, McDade JE (1987) Unexplained febrile illnesses after exposure to ticks. J Am Vet Med Assoc 257:3100–3104
Gossett KA, Gaunt SD, Aja D (1984) Hepatozoonosis and ehrlichiosis in a dog. J Am Anim Hosp Assoc 21:265–267
Greene CE, Harvey JW (1984) Canine ehrlichiosis. In: Greene CE (ed) Clinical microbiology and infectious diseases of the dog and cat. WB Saunders, Philadelphia
Groves MG, Dennis GL, Amyx HL, Huxsoll DL (1975) Transmission of *Ehrlichia canis* to dogs by ticks (*Rhipicephalus sanguineus*). Am J Vet Res 36:937–940
Harvey JW, Simpson CF, Gaskin JM, Sameck JH (1979) Ehrlichiosis in wolves, dogs and wolf-dog crosses. J Am Vet Med Assoc 175:901–905
Hibler SC, Greene CE (1986) Rickettsial infections in dogs Part 11. Ehrlichiosis and infectious cyclic thrombocytopenia. Comp Contin Educ Pract Vet 8:106–114
Hildebrandt PK, Conroy JD, McKee AE, Nyindo MBA, Huxsoll DL (1973a) Ultrastructure of *Ehrlichia canis*. Infect Immun 7:265–271
Hildebrandt PK, Huxsoll DL, Walker JS, Nims RM, Taylor R, Andrews M (1973b) Pathology of canine ehrlichiosis (tropical canine pancytopaenia). Am J Vet Res 34:1309–1320
Huxsoll DL (1976) Canine ehrlichiosis (tropical canine pancytopaenia): a review. Vet Parasitol 2:49–60
Huxsoll DL, Hildebrandt PK, Nims RM, Amyx HL, Ferguson JA (1970a) Epizootiology of tropical canine pancytopaenia. J Wildl Dis 6:220–225
Huxsoll DL, Hildebrandt PK, Nims RM, Walker JS (1970b) Tropical canine pancytopaenia. J Am Vet Med Assoc 157:1672–1682
Huxsoll DL, Amyx HL, Hemelt IE, Hildebrandt PK, Nims RK, Gochenour WS (1972) Laboratory studies of tropoical canine pancytopaenia. Exp Parasitol 31:53–59
Immelman A, Button C (1973) *Ehrlichia canis* infection (tropical canine pancytopaenia or canine rickettsiosis). J S Afr Vet Assoc 44:241–245
Kakoma I, Carson CA, Ristic M, Huxsoll DL, Stephenson EH, Nyindo MBA (1977) Autologous lymphocyte-mediated cytotoxicity against monocytes in canine ehrlichiosis. Am J Vet Res 38:1557–1559
Kakoma I, Carson CA, Ristic M (1980) Direct and indirect lymphocyte participation in the immunity and immunopathology of tropical canine pancytopaenia – a review. Comp Immun Microbiol Infect Dis 3:291–298
Keefe TJ, Holland CJ, Salyer PE, Ristic M (1982) Distribution of *Ehrlichia canis* among military dogs in the world and selected civilian dogs in the United States. J Am Vet Med Assoc 181:236–238
Klopfer U, Nobel TA (1972) Canine ehrlichiosis (tropical canine pancytopaenia) in Israel. Refu Vet 29:24–29
Kuehn NF, Gaunt SD (1985) Clinical and hematologic findings in canine ehrlichiosis. J Am Vet Med Assoc 186:355–358
Leeflang P, Perie NM (1972) A comparative study of the pathogenicity of Old and New World strains of *Ehrlichia canis*. Trop Anim Health Prod 4:107–108
Lewis GE, Huxsoll DL, Ristic M, Johnson AJ (1975) Experimental induced infection of dogs, cats and non-human primates with *Ehrlichia equi*, etiological agent of equine ehrlichiosis. Am J Vet Res 36:85–88
Lewis GE, Ristic M, Smith RD, Lincoln T, Stephenson EH (1977) The brown dog tick (*Rhipicephalus sanguineus*) and the dog as experimental hosts of *Ehrlichia canis*. Am J Vet Res 38:1953–1955
Lovering SL, Pierce KR, Adams LG (1980) Serum complement and blood platelet adhesiveness in acute canine ehrlichiosis. Am J Vet Res 41:1266–1271
Maeda K, Markowitz, N, Hawley R, Ristic M, Cox D, McDade JE (1987) Human infection with *Ehrlichia canis*, a leukocytic rickettsia. N Engl J Med 316:853–856

Matus RE, Leiffer CE, Hurvitz AI (1987) Use of plasmaphoresis and chemotherapy for treatment of monoclonal gammopathy associated with *Ehrlichia canis* infection in a dog. J Am Vet Med Assoc 190:1302–1304

Meinkoth JH, Horer JP, Cowell RL, Tyler RD, Link LJ (1989) Ehrlichiosis in a dog with seizures and non-regenarative anaemia. J Am Vet Med Assoc 195:1754–1755

Mudaliar SV (1944) Canine ehrlichiosis in south India – a preliminary note. Indian Vet J 44:163–164

Neitz WO, Steyn HP (1947) The transmission of *Babesia canis* (Piana and Galli-Valerio, 1895) to the blackbacked jackal (*Thos mesomelas mesomelas* (Schreber)) J S Afr Vet Med Assoc 18:1–15

Neitz WO, Thomas AD (1938) Rickettsiosis in the dog. J S Afr Vet Med Assoc 9:166–174

Nyindo M, Ristic M, Huxsoll DL, Smith AR (1971) Tropical canine pancytopaenia: in vitro cultivation of the causative agent – *Ehrlichia canis*. Am J Vet Res 32:1651–1658

Nyindo M, Huxsoll DL, Ristic M, Kakoma I, Brown JL, Carson CA, Stephenson EH (1980) Cell-mediated and humoral immune responses of German Shepherd dogs and beagles to experimenal infection with *Ehrlichia canis*. Am J Vet Res 41:250–254

Ogunkoya AB, Adeyanju JB, Aliu YO (1981) Experiences with the use of imizol in treating canine blood parasites in Nigeria. J Small Anim Pract 22:775–777

Pienaar U de V (1963) The large mammals of the Kruger National Park – their distribution and present-day status. Koedoe 6:1–37

Pierce KR, Marrs GE, Hightower D (1977) Acute canine ehrlichiosis: platelet survival and factor 3 assay. Am J Vet Res 38:1821–1825

Price JE, Dolan TT (1980) A comparison of the efficacy of imidocarb dipropionate and tetracycline hydrochloride in the treatment of canine ehrlichiosis. Vet Rec 107:275–277

Price JE, Karstad LH (1980) Free-living jackals (*Canis measomelas*) – potential reservoir hosts for *Ehrlichia canis* in Kenya. J Wildl Dis 16:469–473

Price JE, Sayer PD, Dolan TT (1987) Improved clinical approach to the diagnosis of canine ehrlichiosis. Trop Anim Health Prod 19:1–8

Raghavachari K, Reddy AMK (1958) *Rickettsia canis* in Hyderabad. Indian Vet J 35:63–68

Reardon MJ, Pierce KR (1981) Acûte experimental canine ehrlichiosis I. Sequential reaction of the hemic and lymphoreticular systems. Vet Pathol 18:48–61

Ristic M, Huxsoll DL, Weisiger RM, Hildebrandt PK, Nyindo MBA (1972) Serological diagnosis of tropical canine pancytopaenia by indirect immunofluorescence. Infect Immun 6:226–231

Schaer M, Meyer DJ, Young DG (1985) A dual infection of *Leishmania donovani* and *Ehrlichia canis* in a dog. Comp Contin Educ Pract Vet 7:531–534

Seamer J, Snape T (1972) *Ehrlichia canis* and tropical canine pancytopaenia. Res Vet Sci 13:307–314

Simpson CF (1974) Relationship of *Ehrlichia canis* infected mononuclear cells to blood vessels of the lungs. Infect Immun 10:590–596

Smith RD, Ristic M, Huxsoll DL, Baylor RA (1975) Platelet kinetics in canine ehrlichiosis: evidence for increased platelet destruction as the cause of thrombocytopaenia. Infect Immun 11:1216–1221

Smith RD, Sells DM, Stephenson EH, Ristic M, Huxsoll DL (1976) Development of *Ehrlichia canis*, causative agent of canine ehrlichhiosis in the tick *Rhipicephalus sanguineus* and its differentiation from a symbiotic rickettsia. Am J Vet Res 37:119–126

Stephenson EH, Ristic M (1978) Retrospective study of an *Ehrlichia canis* epizootic around Phoenix, Arizona. J Am Vet Med Assoc 172:63–65

Stevenson-Hamilton J (1939) The health of wild animals. J S Afr Vet Med Assoc 10:56–64

Troy GC, Vulgamott JC, Turnwald GH (1980) Canine ehrlichiosis: a retrospective study of 30 naturally occurring cases. J Am Anim Hosp Assoc 16:181–187

Van der Merwe NJ (1959) Die wildehond (*Lycaon pictus*). Koedoe 2:87–93

Van Dijk JE (1971) Studies on *Ehrlichia canis*. Zentralbl Veterinarmed (B) 18:787–803

Van Heerden J (1979) The transmission of canine ehrlichiosis to the wild dog *Lycaon pictus* (Temminck) and black-backed jackal *Canis mesomelas* Schreber. J S Afr Vet Assoc 50:245–248

Van Heerden J (1981) *Ehrlichia canis* besmetting by die hond. M Med Vet (Med) Thesis, University of Pretoria, Pretoria, Republic of South Africa

Van Heerden J (1982) A retrospective study on 120 natural cases of canine ehrlichiosis. J S Afr Assoc 53:17–22

Van Heerden J (1989) Small animal problems in developing countries 1989 In: Ettinger SJ (ed) Textbook of veterinary internal medicine, vol 1. WB Saunders, Philadelphia
Van Heerden J, Goosen DJ (1981) Attempted transmission of canine ehrlichiosis to the vervet monkey (*Cercopithecus pygerythrus*). Onderstepoort J Vet Res 48:127–128
Van Heerden J, Immelman A (1979) The use of doxycycline in the treatment of canine ehrlichiosis. J S Afr Vet Assoc 50:241–244
Van Heerden J, Van Heerden A (1981) Attempted treatment of canine ehrlichiosis with imidocarb dipropionate. J S Afr Vet Assoc 52:173–175
Van Heerden J, Reyers F, Stewart CG (1983) Treatment and thrombocyte levels in experimentally induced canine ehrlichiosis and canine babesiosis. Onderstepoort J Vet Res 50:267–270
Villemin P, Dubor M, Bellangeon M, Fremont Y, Fayette JP (1984) Activite de l'imidocarbe contre differentes protozooses et rickettsioses du chien et du chat. Rev Med Vet 135:441–448
Walker JS, Rundquist JD, Taylor R, Wilson BL, Andrews MR, Barck J, Hogge AL, Huxsoll DL, Hildebrandt RK, Nims RM (1970) Clinical and clinicopathologic findings in tropical canine pancytopaenia. J Am Vet Med Assoc 157:43–55
Weisiger RM, Ristic M, Huxsoll DL (1975) Kinetics of antibody response to *Ehrlichia canis* assayed by the indirect fluorescent antibody method. Am J Vet Res 36:689–694
Willder AG (1977) Prophylactic use of tetracycline for tropical canine pancytopaenia. Vet Rec 101:15

Lyme Borreliosis in Southern Africa

P. Botha[1]

Summary

Lyme borreliosis is transmitted to a variety of animal species including man by infected ticks and biting flies. The illness in humans is characterized by malaise, fever, headache and possibly a rash. Recovery may be spontaneous or patients may develop arthritis, neurological and cardiac manifestations at a later stage. Laboratory diagnosis is fraught with pitfalls. Treatment with beta-lactam antibiotics or tetracycline over several weeks is usually successful. Circumstantial evidence suggests that the disease occurs in South Africa.

1 Introduction

Lyme disease is a newly recognized spirochaetal disease caused by *Borrelia burgdorferi* and is transmitted mainly by ticks. Steere et al. (1977) first reported the findings of an investigation into an outbreak of clustered cases of juvenile arthritis occurring in the Connecticut community of Old Lyme. The mothers of some of these children, questioned the diagnosis of the illness and this led to detailed clinical and epidemiological studies revealing the relationship between tick exposure, at a picnic site and the subsequent development of the disease syndrome. Since ticks and arthritis in humans are both common in southern Africa it seemed logical to look for evidence of the disease in South Africa.

2 The Causative Organism

The spirochaete, *Borrelia burgdorferi* (Johnson et al. 1984; Stanek 1985), is a loose spiral organism 4–30 μm long and 0.18–0.25 μm in diameter and stains with Romanovsky and silver stains (Burgdorfer et al. 1982). There are antigenic differences among strains from different geographical locations and there is mounting evidence that antigenic variation may be a ploy to escape the immune system of the host. Antigens are also shared with other spirochaetes (Barbour et al. 1984; Magnarelli et al. 1987). The nucleic acid composition is enigmatic and comprises minichromosomes or linear plasmids (Hinnebusch and Barbour 1990).

[1]Department of Medical Microbiology, University of the Orange Free State, Bloemfontein 9300, South Africa

2.1 Epidemiology

The disease has been described in North America, Europe and Australia, where the reservoirs are predominantly large ruminants, such as deer, and wild rodents. Migratory birds may act as carriers of infection. The vectors are mainly ticks of the genus *Ixodes* (Spielman et al. 1985). The epidemiology of the disease in humans is related to the life cycle of the vector tick. Transstadial transmission occurs readily while transovarian passage of the *Borrelia* has been observed in *Ixodes pacificus* (Bissett and Hill 1987; Lane and Burgdorfer 1987).

Man may become infected particularly when the nymphae are most active in the summer and human exposure is greatest, while adult ticks are active in autumn and spring. Successful attachment to the host does not ensure infection since transmission has been shown to occur mostly after 48 h of attachment. Frequent searches for ticks and treatment of clothing with pyrethrin would minimize infection by *B. burgdorferi* (Hamilton 1989; Lane 1989).

2.2 Clinical Features

Lyme borreliosis is a multisystem disease which can be divided arbitarily into three recognizable stages. In the earliest stage, a localised skin lesion occurs in most patients at the site of the tick bite within 3–14 days and takes the form of an erythmatous area expanding at the periphery and clearing centrally hence the name erythema chronicum migrans (ECM). Secondary lesions due to hematogenous spread may also appear. Erythema chronicum migrans is pruritic and sensitive and is associated with headache, fever, myalgia, arthralgia and fatigue. The rash regresses spontaneously after about 4 weeks. Less than 20% of these clinical cases may be serologically positive while in endemic areas asymptomatic seropositive individuals may occur.

Transient joint pain is the dominant presenting symptom during the second stage of generalized disease which may occur weeks or months later. The knee joint is most commonly involved. Neurological involvement results in meningitis, facial nerve palsy and radiculopathy (Pachner and Steere 1985; Wilske et al. 1986; Markby 1989; Schmidli et al. 1988).

Progressive arthritis and personality changes dominate the picture in the chronic stage. In a small number of individuals varying degrees of heart block may be observed (Hansen and Madsen 1986; de Koning et al. 1989).

2.3 Diagnosis

Clinical awareness of the syndrome and demonstration of specific IgM and rising convalescent IgG levels are essential criteria in the diagnosis of the disease. The Centers for Disease Control, Atlanta, USA have defined three criteria for the diagnosis of human Lyme disease. The criteria are: (1) previous exposure to ticks; (2) symptoms which include involvement of cardiovascular and nervous system and arthritis; (3) positive serological test (Centers for Disease Control 1984).

2.4 The Immune Response and Diagnostic Tests

The host mounts an IgM and later an IgG response to a number of borrelial antigens (Craft et al. 1984; Magnarelli et al. 1989). A specimen is considered positive using Western blotting if antibodies react with polypeptides having relative mobilities of ca. 31–32, 34–35 and 40 kDa. The American and European strains of *B. burgdorferi* show minor differences and there is growing evidence that the spirochaete is able to evade elimination by the immune system possibly by antigenic variation or sequestering to sites not readily accessible to the immune system (Craft et al. 1986; Anderson et al. 1989).

The indirect fluorescent antibody test (IFAT) can be considered a suitable screening test. Cross-reactions occur with other spirochaetes and in South Africa cross-reactions with *Treponema pallidum* infection may occur frequently. The anti-complement indirect immunofluorescence test is more sensitive and specific (Lane et al. 1990). Enzyme-linked immunosorbent assays (ELISA) may also be used, although at present, the Western blot provides reliable results.

The IFAT proves invaluable in the screening of impression smears of tick tissues in epidemiological studies as only 0.2% of ticks are positive on culture for *Borrelia burgdorferi* (R.S. Lane, pers. comm. 1990).

Rodent tissues may be cultured successfully during the bacteremic phase using appropriate inhibitory antibiotics. Biopsies from the edge of skin lesions may also be submitted for culture (Stanek et al. 1985). Histology of lesions could prove valuable although silver impregnation stains are well known for problematic interpretation.

The spirochaete antigens in the urine of man and animals infected by *B. burgdorferi* may also be detected using monoclonal antibodies (Hyde et al. 1989).

Clinical cases which are sero-negative do occur (Dattwyler et al. 1988). Similarly false positive screening tests may occur in rheumatoid arthritis, autoimmune diseases and other spirochaetal infections (Hamilton 1989). In South Africa leptospirosis, syphilis and relapsing fever due to *B. duttoni* occur and special care should be taken in interpreting serological tests although none of these are likely to cause confusion in the Western blot technique as outlined earlier (R.S. Lane, pers. comm. 1990). In the United States, antibodies to other *Borrelia* are also known to react with the *B. burgdorferi* antigen but due care in interpretation obviates this problem. The antigenic relationship between *B. burgdorferi* and *B. theileri* requires elucidation especially in terms of diagnosing Lyme disease in livestock and feral animals. More recently, the polymerase chain reaction has been reported as a successful aid to diagnosis (Malloy et al. 1990) and investigation is proceeding on this aspect locally.

3 South African Surveys

3.1 Materials and Methods

Sera from human patients in the Bloemfontein area examined for borrelial antibodies fell into two groups. In the first group of more than 500, sera had been submitted for tests other than Lyme borreliosis but the clinical history was suggestive of possible Lyme disease. In the second group, sera were screened from patients who had had a previous history of exposure to ticks. A further 56 sera with specific requests for Lyme serology were also investigated. Three patients with putative disease were residents of the eastern Cape. Sera from 13 horses from the Bloemfontein district, 111 feral rodents from the eastern Cape and one cow were also examined by IFAT. Salivary glands and midgut from 86 ticks, which included *Ixodes*, *Rhipicephalus* and *Otobius* spp. were cultured in Barbour-Stoenner-Kelly Medium (BSK). Serologically positive samples were referred to a reference centre, Hygiene-Institut der Universität in Vienna for immunoblotting. Sera from normal, healthy human subjects were included as controls. When the ELISA antigen became available, ELISA was performed on 36 of those IFAT positive sera[2]. In a separate survey, sera from 85 suspected cases and 233 normal, healthy blood donors in the Johannesburg area were examined for borrelial antibodies by IFAT, (Frean 1989). Western blot was performed on those positive with IFAT.

3.2 Results

No isolations of *B. burgdorferi* were made from tick vectors in the Bloemfontein survey. A large number of false positives were diagnosed in the survey of human sera. In all butone case this was due to positive syphilis serology. One false positive result came from a patient with severe periodontal disease. ELISA produced one discrepant result on the serum of a jaundiced patient with tick bite fever.

In the Johannesburg survey, three sera from possible clinical cases of Lyme disease reacted positively on immunoblotting but only one satisfied the criteria of specificity and this was a visitor from Europe so that it was unclear whether the disease was acquired locally. All the normal controls were negative.

4 The Natal Epidemic

In the early summer of 1988, the owner of a large horse riding school in the Natal province of South Africa became ill, complaining of headaches, an inability to concentrate, fatigue, muscular pains and speech impediment following a single tick bite 10 days previously. Local veterinarians, being alert to the possibility of Lyme disease, because of an outbreak of arthritis in horses in the same riding school (Fivaz et al. 1990) submitted sera from the index case and three other riders from the same stable, who had subsequently developed similar symptoms. Further samples including specimens of the spinose ear tick, *Otobius megnini* found on the horses, sera from stable hands, dogs on the property and all the horses at the riding school were submitted to the Department of Medical

[2] The ELISA antigen was prepared in Vienna as a gift from Dr Stanek.

Microbiology, University of the Orange Free State in Bloemfontein for Lyme serology and bacteriology. Seropositive cases were identified in sera from the 3 riders, the owner, 1 of 4 stable hands, 71 of 117 horses and 5 of 11 dogs[3]. *B. burgdorferi* was not isolated from the ticks. Patients were treated successfully with prolonged courses of intravenously administered penicillin (Botha et al. 1989).

4.1 Treatment

Cases diagnosed early in the disease process should preferably be treated with a penicillin, according to a schedule recommended for syphilis, or alternatively for a prolonged period with tetracycline (Meier et al. 1989). Third generation cephalosporins have been advocated when the disease becomes generalized (Dattwyler et al. 1987; Hamilton 1989). In all cases prolonged treatment is indicated. It has been suggested that the spirochaetes may be sequestered and non-responsive to antibiotic therapy (Hamilton 1989).

4.2 Prevention

Avoiding tick infested areas is important in preventing the disease. Wearing protective garments with or without tick repellant and frequent tick searches further diminish the chances of tick attachment.

5 Discussion

The clinical syndrome following infection by *B. burgdorferi* has led to Lyme disease being called the "new great imitator" (Pachner 1988). Awareness of the situation in which the disease is likely to develop is the first prerequisite which may lead to a diagnosis. The outbreak involving a horse riding stable in Natal is a clear example.

Serological tests must be interpreted with caution. The isolation of *B. burgdorferi* from an infected case is diagnostic but the yield of successful culture from clinical specimens is low.

There are circumstantial indications that Lyme borreliosis occurs in southern Africa and there is an urgent need to isolate the causative organism from the vector tick. The serological differentiation from *B. theileri* and *B. duttoni* must also be established.

Acknowledgements. Grants to support this work were received from the Medical Research Council and the Arthritis Foundation.

---

[3] Confirmed by Dr Gerold Stanek, Hygiene-Institut der Universität, Vienna, Austria.

References

Anderson JF, Magnarelli LA, LeFebvre RB, Andreadis TG, McAninch JB, Perng G-C, Johnson RC (1989) Antigenically variable *Borrelia burgdorferi* isolated from cottontail rabbits and *Ixodes dentatus* in rural and urban areas. J Clin Microbiol 27:13–20

Barbour AG, Tessier SL, Hayes SF (1984) Variation in a major surface protein of Lyme disease spirochaetes. Infect Immun 45:94–100

Bissett ML, Hill W (1987) Characterisation of *Borrelia burgdorferi* strains islolated from *Ixodes pacificus* ticks in California. J Clin Microbiol 25:2296–2301

Botha P, Fivaz BH, Stanek G, McLoed I, Bôhmer L, Hodgkinson BJ (1989) Lyme borreliosis in South Africa (letter). S Afr Med J 76:581

Burgdorfer W, Barbour AG, Hayes SF (1982) Lyme disease – a tick-borne spirochaetosis? Science 216:1317–1319

Centers for Disease Control (1984) Update: Lyme disease – United States. MMWR 33:268–270

Craft JE, Grodzicki RL, Steere AC (1984) Antibody response in Lyme disease: evaluation of diagnostic tests. J Infect Dis 149:789–795

Craft JE, Fischer DK, Shimamoto GT, Steere AC (1986) Antigens of *Borrelia burgdorferi* recognised during Lyme disease. J Clin Invest 78:934–939

Dattwyler RJ, Halperin JJ, Pass H, Luft BJ (1987) Ceftriaxone as effective therapy in refractory Lyme disease. J Infect Dis 155:1322–1324

Dattwyler RJ, Volkman DJ, Luft BJ, Halperin JJ, Thomas J, Golightly MG (1988) Seronegative Lyme disease. N Engl J Med 319:1441–1446

de Koning J, Hoogkamp-Korstanje JAA, van der Linde MR, Crijns HJGM (1989) Demonstration of spirochaetes in cardiac biopsies of patients with Lyme disease. J Infect Dis:150–153

Fivaz BH, Botha P, Cairns LM (1990) A putative outbreak of equine Lyme borreliosis in Natal. S Afr Vet J 61:128–129

Frean J (1989) The identification of Lyme disease in South Africa. Research report submitted to University of the Witwatersrand in partial fulfillment for M Med Path (Microbiol)

Hamilton DR (1989) Lyme disease. Hidden pandemic or media event? Infect Dis Newsl 8:65–69

Hansen K, Madsen JK (1986) Myocarditis associated with tick-borne *Borrelia burgdorferi* infection. Lancet June 7 1986:1323–1324

Hinnesbusch J, Barbour AG (1990) Molecular analysis of linear plasmid telomeres of *Borrelia burgdorferi*. Abstr H 102 90th Anniversary Meeting of American Society for Microbiology 13–17 May 1990 Anaheim, California, USA

Hyde FW, Johnson RC, White TJ, Shelburne CE (1989) Detection of antigen in urine of mice and humans infected with *Borrelia burgdorferi*, etiologic agent of Lyme disease. J Clin Microbiol 27:58–61

Johnson RC, Schmid GP, Hyde FN, Steigerwalt AG, Brenner DI (1984) *Borrelia burgdorferi* sp. nov.; etiological agent of Lyme disease. Int J Syst Bacteriol 34:496–497

Lane RS (1989) Treatment of clothing with permethrin spray for personal protection against *Ixodes pacificus*. Exp Appl Acarol 6:343–335

Lane RS, Burgdorfer W (1987) Transovarial and transstadial passage of *Borrelia burgdorferi* in the western black-legged tick *Ixodes pacificus* (Acari: Ixodidae). Am J Trop Med Hyg 37:188–192

Lane RS, Lenette ET, Madigan JE (1990) Interlaboratory and intralaboratory comparisons of indirect immunofluorescence assays for serodiagnosis of Lyme disease. J Clin Microbiol 28:1774–1779

Magmarelli LA, Anderson JF, Barbour AG (1986) The etiologic agent of Lyme disease in deer flies, horse flies and mosquitoes. J Infect Dis 154:355–358

Magnarelli LA, Anderson JF, Johnson RJ (1987) Cross-reactivity in serological tests for Lyme disease and other spirochaetal infections. J Infect Dis 156:183–188

Magnarelli LA, Anderson JF, Barbour AG (1989) Enzyme-linked immunosorbent assay for Lyme disease: reactivity of subunits of *B. burgdorferi*. J Infect Dis 159:43–49

Malloy DC, Nauman RK, Paxton H (1990) Detection of *Borrelia burgdorferi* using the polymerase chain reaction. J Clin Microbiol 28:1089–1093

Markby DP (1989) Lyme disease facial palsy: differentiation from Bell's palsy. Br Med J 299:605–606

Meier C, Reuben HJ, Huber P, Mumenthaler M (1989) Meningo-radiculitis mimicing vertebral disc herniation. A "neurological" complication of Lyme borreliosis. Acta Neurochir 98:42–46
Pachner AR (1988) *B. burgdorferi* in the nervous system: the "new great imitator". Ann Acad Sci 539:56–64
Pachner AR, Steere AC (1985) The triad of neurological manifestations of Lyme disease : meningitis, cranial neuritis and radiculitis. Neurology 35:47–53
Schmidli J, Hunziken T, Moesli P, Schaad UB (1988) Cultivation of *Borrelia burgdorferi* from joint fluid three months after treatment of facial palsy due to Lyme borreliosis. J Infect Dis 158:905–906
Spielman A, Wilson ML, Levine JF, Piesman J (1985) Ecology of *Ixodes dammini* – borne human babesiosis and Lyme disease. Ann Rev Entomol 30:439–460
Stanek G (1985) Lyme disease and related disorders. Microbiol Sci 2:231–234
Stanek G, Wewelka G, Groh V, Neumann R (1985) Isolation of spirochaetes from the skin of patients with erythema chronicum migrans in Austria. Zentralbl Bakteriol Hyg A 260:88–90
Steere AC, Malawista SE, Snydman DR, Shope RE, Andiman WA, Ross MR, Steele FM (1977) Lyme arthritis: an epidemic of oligoarticular arthritis in children and adults in three Connecticut communities. Arthritis Rheum 20:7–17
Wilske B, Schierz G, Preac-Mursic V, von Busch K, Kuhbeck R, Pfister H-W, Einhaupt K (1986) Intrathecal production of specific antibodies against *Borrelia burgdorferi* in patients with lymphocytic meningoradiculitis (Bannwarth's syndrome). J Infect Dis 153:304–314

Tick-Bite Fever (Tick Typhus) in Southern Africa

J.H.S. GEAR[1]

Summary

Tick-bite fever or tick typhus occurring in southern Africa is a rickettsial disease caused by *Rickettsia conorii* var. *pijperi* transmitted by ixodid ticks. The infection passes from one generation of ticks through the eggs to the next. The larvae, nymphs and adults are potentially infective to man. The infection is transmitted through the bite during feeding. The bite mark develops into an inflamed red papule, the centre of which becomes necrotic and black to form the characteristic primary lesion. The regional lymph glands become enlarged and tender. Systemic signs include chills, muscle and joint pains, headache, and delirium in severe cases. In untreated cases, the fever typically lasts 10 days. In severe cases the patient may lapse into coma and on recovery may have defective speech. Other severely ill patients develop a hemorrhagic state which may be fatal. Diagnosis of tick-bite fever may be confirmed in the laboratory by the isolation and identification of the rickettsiae or by specific serological tests. The illness responds specifically to treatment with chloramphenicol and tetracycline antibiotics which usually results in a dramatic improvement after 48 h. The striped mouse *Rhabdomys pumilio* and the vlei rat *Otomys irroratus* may serve as reservoirs of infection. The domestic dog and the common black rat, *Rattus rattus*, may acquire the infection and may be important sources of infection to humans. Tick-bite fever may be avoided by avoiding contact with ticks. Protective vaccines have also been produced experimentally but are not in general use.

1 Rickettsiae and Disease

Tick-bite fever is a rickettsial disease. The rickettsiae are small bacteria-like, obligate, intracellular parasites of arthropods. Some of them, when transmitted to man, cause disease. Amongst the diseases so caused are the typhus group of fevers which may be grouped according to the vector into epidemic louse-borne typhus fever caused by *Rickettsia prowazeki*, murine flea-borne typhus fever caused by *Rickettsia mooseri*, and tick-borne typhus fever, in the Eastern Hemisphere caused by *Rickettsia conori* and in the Western Hemisphere called Rocky Mountain spotted fever and caused by *Rickettsia rickettsi*.

Another rickettsial disease transmitted by ticks is Q-fever, caused by *Rickettsia burneti*, also known as *Coxiella burneti*. Q-fever, because of certain biological differences of the causative organism and differences in the clinical manifestations of the disease, is not classified with the typhus group of fevers.

The tick typhus group of fevers is divided into two forms:

1. Rocky Mountain spotted fever, occurring in the Western Hemisphere.

---

[1]National Institute for Virology, Private Bag X4, Sandringham 2131, South Africa

2. Tick typhus fever of the Eastern Hemisphere including fievre boutonneuse, also known as Mediterranean spotted fever, Indian and Siberian tick typhus, Queensland tick typhus and Kenyan tick typhus.

The rickettsiae causing tick typhus in the Eastern Hemisphere are all closely related to one another. The form occurring in southern Africa is caused by *Rickettsia conori* var. *pijperi*, and is transmitted by various species of ixodid ticks including species of the genera *Rhipicephalus*, *Amblyomma*, *Hyalomma*, *Boophilus* and *Haemaphysalis* (Gear 1954).

1.1 Transmission

The infection is transmitted transstadially and transovarially. The larvae, nymphs and adults are potentially infective to man. However, most human infections are transmitted by larval ticks which are small and are not readily detected by human hosts. Larvae are also not necessarily host-specific and will readily feed on man as well as their more natural hosts. Nymphs and adults are more host specific, are larger and are more readily detected on the skin and are usually removed before attaching. These two instars are not often responsible for transmission to human beings, although they are involved in some cases (Gear and de Millon 1941).

2 Epidemiology of Tick-Bite Fever in Southern Africa

Tick-bite fever and related forms of tick typhus are widespread in sub-Saharan Africa. In South Africa, tick-bite fever occurs in every region except in the semi-desert regions of the western part of southern Africa where the arid conditions are inimical to ticks. The infection is especially common in bushveld areas which provide a favourable habitat for large numbers of ticks. The ticks concerned in the transmission are species of most genera of the Ixodidae including species of *Rhipicephalus*, *Amblyomma*, and *Hyalomma*. Urban infection occurs through the bites of the dog ticks, in particular the yellow dog tick *Haemaphysalis leachi*, and the kennel tick *Rhipicephalus sanguineus*. It has been shown experimentally that the striped mouse, *Rhabdomys pumilio*, and the vlei rat, *Otomys irroratus* are often infected in nature and may serve as a source of infection to previously uninfected ticks (Gear, unpubl. observ.). The common black rat, *Rattus rattus*, living in the suburbs has also been shown to be infected. Dogs may actually suffer from the infection and manifest a characteristic bite mark, lymphadenitis and fever, and presumably they may assist in maintaining urban infections (Gear, unpubl. observ.). The infection passes from one generation of ticks through the eggs to the next, each stage of which, the larva, the nymph and the adult is infected and potentially infective while feeding. This mode of transmission has been shown to continue for five generations and presumably may continue indefinitely (Gear 1978). The natural history of *R. conori* is summarized in Fig. 1.

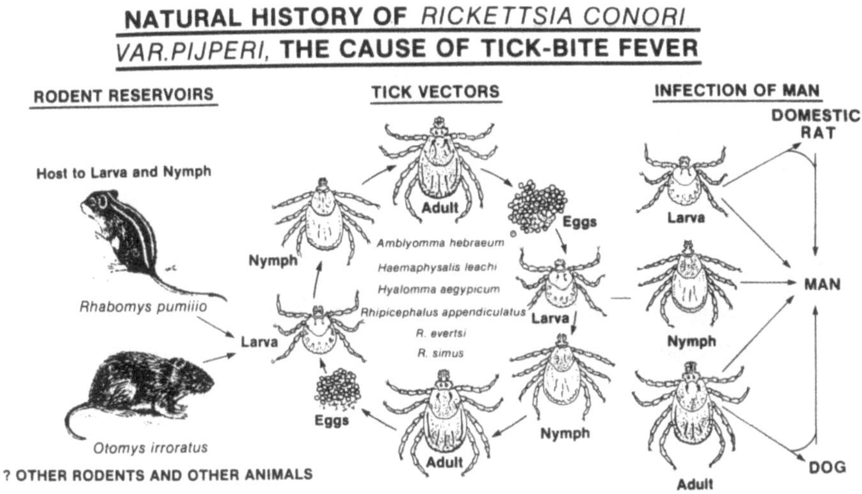

Fig. 1. Natural history of *Rickettsia conori* var. *pijperi*, the cause of tick-bite fever. See comment on "Hyalomma aegyptium" on p. 145

Soft ticks (Argasidae) do not play an important role in the transmission of tick-bite fever, but a rare case has come to attention where they were responsible for transmission to man. The common human tampan tick, *Ornithodoros moubata*, was used to transmit the rickettsiae of tick-bite fever from South Africa to the USA for experimental studies during World War II. The tampan ticks were fed on infected guinea pigs during the first two days of their fever until engorged, then suitably packed and transmitted by post to the Rocky Mountain spotted fever laboratory. Six months elapsed before they arrived and on arrival the strain of tick-bite fever was successfully established in guinea pigs and then in egg culture. A strain of *Rickettsia prowazeki*, the cause of epidemic louse-borne typhus fever, was similarly sent to USA.

2.1 The Organism

The rickettsiae causing tick-bite fever are pleomorphic organisms frequently seen in pairs surrounded by a halo as if encapsulated. They average about 1.2×0.6 µm. Electron micrographs have revealed that the organism has a limiting membrane similar to the wall of bacteria which contains muramic acid. Biochemical studies have revealed that the rickettsia contains both DNA and RNA. The rickettsiae, therefore, are more closely related to bacteria than to viruses and apparently multiply by binary fission. They are relatively labile organisms, readily inactivated by heat and chemical agents. They are rendered non-infectious within 24 h by 0.1% formaldehyde or 0.5% phenol. They remain viable for long periods when stored at $-70\,°C$ in whole tissue or in protective media. When lyophilized on protein media and sealed under nitrogen and stored at 5 °C they remain viable for a number of years.

2.2 Culture

R. conori have been grown in a number of tissue culture systems, however chick embryos are routinely used for their propogation. Guinea pigs have also been routinely used for their isolation and study. After intraperitoneal inoculation of blood from patients suffering from tick-bite fever, a guinea pig develops a rectal temperature of over 40 °C after an incubation period of 4–7 days in contrast to its normal rectal temperature of 39 °C. Strains so isolated can be stabilized in guinea pigs by passage of peritoneal washings collected on the first or second day of fever. During the infection adult male guinea pigs develop a scrotal reaction with inflammatory swelling and edema of the scrotal sac. Guinea pigs have also been used to characterize rickettsial strains using cross-immunity tests. Thus guinea pigs recovered from infection with strains of tick-bite fever, when challenged with other strains of homologous tick-bite fever rickettsiae, are solidly immune. However, recent findings have suggested that there are antigenic differences between field and suburban strains of rickettsiae. Guinea pigs which have recovered from tick-bite fever are also solidly immune when challenged with strains of fievre boutonneuse and vice versa. Guinea pigs resistant to *R. conori* show a partial immunity to *R. rickettsi* and vice versa.

Microscopic examination of impression smears made from the peritoneal surface of infected guinea pigs, gerbils, or from the yolk sac membrane, reveals a characteristic pattern of infection with the rickettsiae occurring throughout the cytoplasm and singly or in small clumps in the nuclei of cells. This pattern differs from that of louse-borne and flea-borne typhus in which the rickettsiae occur massed together, filling or almost filling the cytoplasm of the infected cells.

2.3 Pathogenesis and Pathology

The infection is transmitted to man through the tick bite during feeding. Occasionally it may be transmitted via the ocular conjunctiva after handling infected tick material.

In the skin, the reaction to the local multiplication of rickettsiae at the site of the bite leads to the formation of a raised, red papule. The inflammation and thrombosis of the affected capillaries leads in turn to necrosis of the centre of the papule and the formation of a typical raised red lesion with a black centre known as the "tache noir". When the infection is introduced through the conjunctiva of the eye, marked inflammation with dilation of the capillaries associated with swelling and edema of the tissue of the eyelids occurs. The swelling may be so marked as to close the eye, a condition known as chemosis. In addition to marked congestion, shallow ulcers may form on the conjunctiva.

The rickettsiae then flow from the primary site of infection with the lymph to the regional lymph glands which become enlarged. Microscopic examination of the lymph nodes may show rickettsiae within the endothelial cells associated with an infiltration of inflammatory cells, mostly mononuclear cells, and a few neutrophil leucocytes. The infection then spreads into the general circulation and the resulting rickettsaemia leads to the establishment of numerous foci of infected cells in the blood vascular system. Platelets and white cells, especially mononuclear cells adhere to the vascular surface of the infected endothelial cells which is accompanied by a perivascular infiltration of inflammatory cells including mononuclear cells, histiocytes and neutrophil leucocytes. The whole lesion constitutes a "typhus node" which is the unit lesion of the infection in the typhus group of fevers. In tick typhus, the typhus nodes are relatively larger than those of epidemic louse-borne typhus fever due to *R. prowazeki*. These nodes are found in the skin and their presence manifests with the characteristic rash of tick-bite fever, in which the individual elements tend to be coarser than those observed in epidemic typhus fever. In fatal cases, numerous typhus nodes are seen in the brain and their presence and associated inflammation and edema is responsible for the intense headache, delirium, stupor and coma which may occur in severe cases of tick-bite fever. Hemorrhaging may occur in some patients with severe tick-bite fever. This is associated with a marked thrombocytopenia which results in bleeding from the gastrointestinal and respiratory mucous membranes. These cases are liable to be fatal (Walker and Gear 1986).

3 Symptoms

Many patients with tick-bite fever tell of visiting the bushveld or the coastal strip of South Africa where ticks abound. Other patients acquire the infection while on visits to farms where cattle and other large domestic animals frequently harbour many ticks. In the suburbs of the cities and towns, the infection may be transmitted by dog ticks which may be carried into the houses by domestic pets which are often allowed to sleep on the patient's bed or in babies' cots.

The incubation period from the time of infection to the onset of general symptoms is about 1 week. Some patients complain of numerous irritating tick bites but often the patient is unaware of the infecting bite. During this incubation period one or more of the tick bites develop into an inflamed red papule, the centre of which becomes necrotic and black to form the characteristic primary sore or "tache noir". This may be situated anywhere on the surface of the body. In infants and babies it is often on the scalp, where it may be difficult to detect because of the hair. In adults and older children the bite is often found on the lower limbs, in the groin or on the abdomen. The lymph glands draining the area of the bite become enlarged and tender but the bite mark itself is usually painless. The patient may be aware of enlarged tender lymph glands but unaware of the bite lesion. The "tache noir" is the most typical sign of tick-bite fever and is present in most patients. However, in some it cannot be found. If the infection occurs through the conjunctiva the preauricular lymph gland draining the eye tissues becomes enlarged and tender and this is followed by symptoms of systemic infection.

3.1 Systemic Manifestations of Tick-Bite Fever

The patient may initially complain of lassitude which increases in the evening but the following morning he/she may feel better but deteriorates in the evening and complains of chilliness, slight shivers, muscle and joint pains and headache.

The fever reaches its peak on the second or third day and in untreated cases continues for 10 days. The febrile reaction shows an intermittent or remittent course. In mild cases the fever may last from 1–7 days. In severe cases, it may continue with slight remissions for 14 days or even longer. The outstanding symptom of tick-bite fever is headache, often very severe, and the patient may become delirious and may have visual hallucinations. In very severe cases the patient may become stuporose and then comatose.

On examination, the characteristic "tache noir" and regional lymphadenitis may be noted. The face tends to be flushed and the conjunctivae injected. The tongue may be slightly coated and the throat slightly inflamed with slight general lymphadenopathy. The heart rate is not greatly increased, the abdomen not distended and the spleen and liver not palpable. There may be some tenderness in both upper quadrants of the abdomen. A characteristic maculopapular rash erupts on the third to the fifth day of illness, appearing first on the extremities and then on the trunk. Fresh macules and papules may be noted for 1–3 days. Papules can be felt as small circumscribed nodules in the skin, at first pinkish, later becoming darker. Characteristically the rash involves the palms of the hands, the soles of the feet, and to a lesser extent the face. The profuseness of the rash is directly related to the severity of the illness. In mild cases only a few papules may be detected. In severe cases a profuse maculopapular rash involves the whole body, usually more marked peripherally than centrally on a dark cyanotic skin. In very severe cases the rash may become hemorrhagic (Gear 1954).

3.2 Course and Complications

Tick-bite fever in children and young adults is usually a mild illness and they make an uneventful recovery. The illness tends to be severe in middle aged and elderly patients and, if not treated early, the patient may develop serious complications. Most complications involve the blood vessels, the most frequent being deep vein thrombosis of the leg late in the illness or early in convalescence and may be further complicated by pulmonary embolism which may be fatal. Other veins may be involved. When the retinal veins are affected there is serious interference with vision. The retina may also be involved by a typhus node resulting in defective vision of the affected eye, fortunately, usually without permanent damage and normal vision is restored during convalescence (Gear 1939).

In mild cases little change in the blood pressure is noted but in severe cases there may be a significant fall with poor circulation, sometimes leading to gangrene of the fingers and toes. Myocarditis may occur, resulting in fatal circulatory collapse.

Serious involvement of the brain may result in stupor and coma, and on recovering consciousness serious interference with neurological function is apparent including loss of speech, a complication noted recently in three young children (Gear et al. 1990). Perhaps the most serious complication is the development of a hemorrhagic crisis manifesting as epistaxis, hematemesis and melena, occasionally with a fatal outcome (Gear et al. 1983).

4 Diagnosis

The diagnosis of tick-bite fever is usually based on clinical grounds alone with the finding of the primary tick bite lesion, regional lymphadenitis and the characteristic maculopapular rash. It may be specifically confirmed by demonstrating the presence of *R. conori* in skin biopsies of the primary lesion or the papular elements of the rash using immunofluorescence. The test thus provides one of the earliest laboratory clues to the nature of the patient's illness. *Rickettsia conori* may be isolated by the intraperitoneal inoculation of the patient's blood into guinea pigs which subsequently develop typical febrile, non-fatal illness. The strain may be established by successive passage in guinea pigs on the first or second day of fever. The rickettsiae may be seen showing the typical cytological picture in suitably stained impression smears made from the peritoneal surface of the spleen or testes. Guinea pigs which survive may be bled 4–6 weeks after inoculation and the acute and convalescent phase sera tested for antibodies against *R. conori*. In practice the diagnosis usually depends on serological tests to demonstrate the appearance of antibodies in the patient's blood using agglutination, complement fixation and immunofluorescence (Gear 1978).

5 Treatment

Tick-bite fever responds specifically to treatment with chloramphenicol or tetracycline antibiotics.

The initial recommended dose of chloramphenicol is 50 mg/kg and of tetracycline 25 mg/kg followed by the same amount 8 hourly and continued for one day after defervescence. Generally little change in the patient's condition occurs during the first 24 h but there is a dramatic improvement after 48 h. The rash fades rapidly and the patient feels better. In severely ill patients the response may take longer than the usual 48 h but recovery usually takes place even in the most severe cases. Chloramphenicol is a suitable alternative to tetracycline in patients for whom tetracycline is contra-indicated.

The illness may be prevented by initiating treatment as soon as the appearance of the typical tick-bite and regional lymphadenitis suggest that the patient has become infected with *R. conori*. Patients treated in this way develop little immunity and are susceptible to another attack of tick-bite fever when again exposed to the bites of infected ticks. Complications require the appropriate urgent attention and therapy.

6 Prevention

Tick-bite fever may be avoided by preventing contact with ticks. This may be achieved by avoiding rural areas and walking through grass or bush unless appropriate preventative measures are taken. In suburbs where the infection is usually transmitted by dog ticks, it has been found that families who allow their pet dogs to come indoors regularly experience cases of tick-bite fever and therefore this practice should be avoided. Manual removal of ticks from domestic pets should be discouraged.

The application of pyrethroid acaricide sprays and tick repellants such as dimethyl phthalate applied to socks and the lower end of trousers and stocking may help in preventing ticks from attaching. A careful inspection for attached ticks following exposure may also be of value in preventing tick-bite fever. In practice, the risk of tick-bite fever is readily accepted by most visitors to rural areas, in the knowledge that, if they acquire the infection it can be speedily and effectively treated. Untreated cases may be fatal however. The importance of early diagnosis cannot be overemphasized.

A vaccine against tick-bite fever was produced in large amounts at the South African Institute for Medical Research during World War II to immunize the soldiers operating in the bushveld areas of South Africa. It was prepared successfully from egg cultures of the rickettsiae in the yolk sac membrane. The suspension of the rickettsiae was purified and refined by extraction with ether which removed the yolk and cell debris leaving a pure suspension of rickettsiae which was inactivated by the addition of formalin. In experimental studies this vaccine was found to have a protective effect but it was not possible to determine its value in humans. As highly effective treatment for tick-bite fever is now available, the need for the vaccine has largely fallen away but it may be of value in protecting individuals who frequently encounter ticks in their work or recreation (Gear 1969).

References

Gear JHS (1939) Complications of tick-bite fever – a survey of fifty cases. S Afr Med J 13:35–36

Gear JHS (1954) The rickettsial diseases of South Africa – a review of recent studies. S Afr J Clin Sci 5:158–175

Gear JHS (1969) Rickettsial vaccines. Br Med Bull 25:171–176

Gear JHS (1978) Tick-bite fever. S Afr J Hosp Med 4:45–49

Gear JHS, de Meillon B (1941) Hereditary transmission of the rickettsiae of tick-bite fever through the common dog tick *Haemaphysalis leachi*. S Afr Med J 15:389–392

Gear JHS, Miller GB, Martins H, Swanepoel R, Wolstenholme B, Coppin A (1983) Tick-bite fever in South Africa; the occurrence of severe cases on the Witwatersrand. S Afr Med J 73:807–810

Gear JHS, Wagner JM, Dyssell JCG, Hutton SA, Wehde SDG (1990) Severe tick-bite fever in young children. S Afr Med J 77: 422–424

Walker DH; Gear JHS (1986) Correlation of the distribution of *Rickettsia conorii* microscopic lesions and clinical features in South African tick-bite fever. Am J Trop Med Hyg 34:361–371

Observations on African Tick Typhus (Tick-Bite Fever) in Zimbabwe

C.E. Yunker[1] and R.A.I. Norval[1]

Summary

African tick typhus, caused by *Rickettsia conori*, affects humans throughout most of sub-Saharan Africa. Widely distributed in Zimbabwe, it is particularly prevalent in the southern lowveld, where numerous cases were seen in troops during the civil unrest of the late 1970s. Recent experiences of visitors to our Heartwater Research Station in this area have confirmed its continuing prevalence. In view of recent reports of severely debilitating and sometimes fatal infections with *R. conori* throughout its range, as well as a tendancy by local physicians to treat the infection inadequately, we briefly review the diagnosis, treatment and vectors of the disease in Zimbabwe.

1 Introduction

African tick typhus (tick-bite fever), caused by *Rickettsia conori* and transmitted to humans by a variety of ixodid ticks, is endemic throughout most of the Ethiopian faunal region, including Zimbabwe. It is generally considered to be a mild and abortive febrile illness, distinguished by a primary sore (eschar) and secondary *Pymphadenitis*. However, severely debilitating, even fatal, infections have been reported in southern Africa (Gear et al. 1983, 1990; Walker and Gear 1985). Likewise, the Mediterranean variety of this disease, boutonneuse fever, has increasingly been shown to occur in malignant form with viscerotropic involvement and significant fatal outcome (Raoult et al. 1986; Walker et al. 1986). In both varieties of the disease, older patients, particularly males, are more at risk of severe illness, as are those with glucose-6-phosphate dehydrogenase deficiency or suffering from alcohol abuse (Gear et al. 1983; Walker et al. 1986).

2 The Situation in Zimbabwe

We wish to draw attention to the prevalence of tick typhus in Zimbabwe and the possibility of serious complications if inadequately treated. Within the past three years, nine visitors to our Heartwater Project field station at Mbizi and nearby areas of the southern lowveld have contracted the illness following tick-bite. The infections all responded promptly to treatment with tetracyclines, which, in some cases, were taken despite the reluctance of attending physicians to

[1] University of Florida/USAID/SADCC Heartwater Research Project, c/o Veterinary Research Laboratory, PO Box 8101, Causeway, Zimbabwe

prescribe this antirickettsial agent. Four physicians attempted to treat with inappropriate antibiotics or drugs such as penicillin, erythromycin or sulfonamides. The first two medications are ineffective and the last, which enhances ricksettsial growth in vitro, is contra-indicated (Brezina and Kazan 1989). In two cases thus treated, the patients experienced protracted fever, malaise and muscular soreness, which dramatically abated when tetracyclines were substituted for the ineffective medications. In a separate instance, the physician, although aware of the drug of choice for rickettsial infections, recommended that it be withheld for some days to permit the development of immunity. Another practitioner expressed a preference for serological evidence of infection before prescribing treatment. However, no routine technique is yet available with which to consistently and accurately diagnose a rickettsial etiology early enough to insure prompt specific therapy (McDade et al. 1981). Serology, while useful in confirming a suspected rickettsial infection, is of little value early on. Although *R. conori* has been specifically identified by immunofluorescence test in skin biopsies taken from early boutonneuse fever cases (Raoult et al. 1986) few laboratories located within the distributional range of African tick typhus are capable of isolating and identifying the causal organism by this or any other means. Thus, diagnosis is based on clinical findings and a history of recent tick exposure.

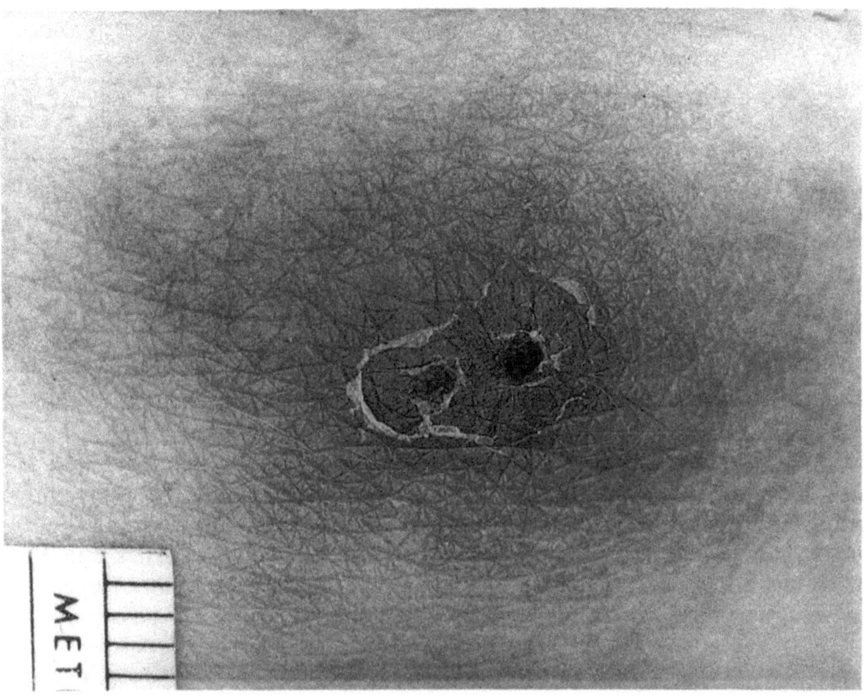

Fig. 1. Primary eschar of tick typhus, approximately 1 week after exposure (scale: 1 gradation = 1 mm)

2.1 Signs

Typical signs include a characteristic eschar with black necrotic core at the site of tick bite (often on the axilla, perineum, leg or scalp margin) (Fig. 1), regional lymphadenitis, fever, headache, malaise, myalgia and neck stiffness. A generalized maculopapular rash usually appears on the fifth day after onset. Characteristically it involves the palmar and plantar surfaces, and face, and tends to be more profuse in severe cases (Gear 1978).

2.2 Treatment

Tetracyclines are the drugs of choice for treatment of rickettsial diseases (Kucers and Bennett 1981). Orally administered oxytetracycline (50 mg/kg per day for 7 days) has consistently proven effective in treating children ill with boutonneuse fever (Moraga et al. 1982). Severe cases in adults have been resolved by administration of 500 mg of the drug at 6-h intervals, and milder cases responded to a lower dosage (Gear 1978; Gear et al. 1983). In all cases, treatment should continue for 24–48 h after defervescence. Choramphenicol is also effective but toxic. It is preferred over tetracyclines for very severe infections, particularly if parenteral administration is indicated or renal failure is involved (Kucers and Bennett 1981).

3 Occurrence

Throughout its vast range (S Europe, Africa, Asia Minor, Crimea, India and SE Asia) *R. conori* has been detected in numerous species representing seven genera of Ixodidae (Hoogstraal 1967; Camicas 1975). In South Africa alone, the rickettsia has been isolated from at least five species of ticks: *Amblyomma hebraeum* Koch, 1844; *Rhipicephalus appendiculatus* Neumann, 1901; *R. evertsi* Neumann, 1897; *Haemaphyalis leachi* (Audouin, 1827) and *Hyalomma* "*aegyptium*"[2], prompting the belief that most ixodids are capable of transmitting it (Gear 1954). In Zimbabwe, as in South Africa (Hoogstraal 1956; Gear 1978), adequate evidence exists to incriminate the bont tick, *A. hebraeum*, a feral species, as the major vector of the disease. During the civil unrest, that occurred in Rhodesia during the late 1970s, tick control on cattle ceased on the tribal areas of the southern lowveld where this species occurs. The abundance of ticks increased and several thousand cases of tick typhus were reported among military personnel. Army medical officers collected ticks from approximately 100 affected soldiers and submitted them to one of us (R.A.I.N.) for identification. All were larvae of *A. hebraeum* (unpublished information). Ticks removed

[2] *Hyalomma aegyptium* (Linne 1758) does not occur in South Africa. The tick yielding *R. conori* may have been *H. marginatum rufipes* Koch, 1844 or *H. truncatum* Koch 1844, inasmuch as the name has been used, in part, for both species (J. Walker, pers. comm.).

from persons visiting our lowveld field station have also been mainly *A. hebraeum*. In the northern lowveld of the Zambezi valley, *A. variegatum* (Fabricius 1794) is undoubtedly involved; elsewhere in Africa *R. conori* has been isolated from this species on a number of occasions (Camicas 1975).

In the Zimbabwean highveld, the main feral vectors are larvae and nymphs of the brown ear tick *Rhipicephalus appendiculatus*, which are commonly found in small game parks. Peridomestic vectors (Norval 1983) are the yellow dog tick, *Haemaphysalis leachi*, and the kennel tick, *R. sanguineus* (Latreille 1806). It is reported that "suburban" cases tend to be more severe than ferally acquired infections (Gear et al. 1983), but the possibility that the former may involve a higher proportion of middle-aged and elderly persons has not been excluded.

3.1 Variations Among Isolates or Strains

Recently, significant antigenic differences were revealed by immunoflourescence tests between Zimbabwean isolates of spotted fever group organisms from *Amblyomma hebraeum* on cattle and those from other ticks (*Rhipicephalus simus* and *Haemaphysalis leachi*) on Zimbabwean dogs. The former isolates, but not the latter, were also markedly different from the reference strain, Simko, of *R. conori* (Kelly and Mason 1990). In that study, differences in severity of clinical response to infection were seen in guinea pigs inoculated with ticks from the two different hosts. It was suggested that: (1) a spectrum of serologically related strains of tick-typhus organisms exist in Zimbabwe and (2) previously described differences in clinical forms of human infections may correlate with the serotype of the etiological agent.

Regardless of serotype, it is emphasised that, if tick typhus is suspected, specific treatment should be initiated without delay in order to avoid pathologic consequences (Gear et al. 1983, 1990; Walker and Gear 1985).

Acknowledgement. This contribution is supported in part by the United States Agency for International Development (Contract No. AFR-0000-C-00-6003-00 and Cooperative Agreement No. AFR-0435-A-00-9084-00).

References

Brezina R, Kazan J (1989) Ch. 5, Rickettsial diseases. In: Goldsmith R, Heyneman D (eds) Tropical medicine and medical parasitology. Appleton and Lange, E Norwalk, CT, USA

Camicas J-L (1975) Conceptions actuelles sur l'épidémiologie de la Fièvre Boutonneuse dan la Région Ethiopienne et la Sous-Région Europeenne Méditerraneenne. Cah ORSTOM Ser Entomol Med Parasitol 13:229–232

Gear JHS (1954) The rickettsial diseases of South Africa. A review of recent studies. S Afr J Clin Sci 5:158–175

Gear JHS (1978) Tick-bite fever. S Afr J Hosp Med 4:45–49

Gear JHS, Miller GB, Martins H, Swanepoel R, Wolstenholme B, Coppin A (1983) Tick-bite fever in South Africa. The occurrence of severe cases on the Witwatersrand. S Afr Med J 63:807–810

Gear JHS, Wagner JM, Dyssell JCG, Hulton SA, Wehde SDG (1990) Severe tick-bite fever in young children. A report of 3 cases. S Afr Med J 77:422–424
Hoogstraal H (1956) African Ixodoidea I. Ticks of the Sudan (with special reference to Equatoria Province and with preliminary reviews of the genera *Boophilus*, *Margaropus* and *Hyalomma*). Res Rep NM 005 050.29.07 U.S. Govt, Dept of the Navy, Bur Med Surg, Washington, DC, pp 1101
Hoogstraal H (1967) Ticks in relation to human diseases caused by *Rickettsia* species. Annu Rev Entomol 12:377–420
Kelly PJ, Mason PR (1990) Serological typing of spotted fever group rickettsia isolates from Zimbabwe. J Clin Microbiol 28:2302–2304
Kucers A, Bennett NMcK (1981) The use of antibiotics. 3rd edn Wm. Heinemann Medical Books, London, pp 441, 624–625
McDade JE, Wells DE, Brooks JB, Alley CC (1981) Gas-liquid chromatography of acute-phase serum as a technique for early diagnosis of rickettsial diseases. In: Burgdorfer W, Anacker RL (eds) Rickettsiae and Rickettsial Diseases. Academic Press, New York, pp 631–639
Moraga FA, Martinez-Roig A, Alonso JL, Boronat M, Domingo F (1982) Boutonneuse fever. Arch Dis Child 57:149–151
Norval RAI (1983) The ticks of Zimbabwe. VII. The genus *Amblyomma*. Zimbabwe Vet J 14:6–18
Raoult D, Weiller PJ, Chagnon A, Chaudet H, Gallais H, Casanova P (1986) Mediterranean spotted fever: clinical laboratory and epidemiological features of 199 cases. Am J Trop Med Hyg 35:845–850
Walker DH, Gear JHS (1985) Correlation of the distribution of *Rickettsia conorii*, microscopic lesions, and clinical features in South African tick bite fever. Am J Trop Med Hyg 34:361–371
Walker DH, Staiti A, Mansueto S, Tringali G (1986) Frequent occurrence of hepatic lesions in boutonneuse fever. Acta Trop 43:175–181

The Role of Ticks in the Epidemiology of Crimean-Congo Hemorrhagic Fever in Southern Africa

Y. RECHAV[1]

Summary

The widespread distribution of *Hyalomma marginatum rufipes*, *Hyalomma truncatum* and *Rhipicephalus evertsi evertsi* and the presence of antibodies of CCHF in so many animals suggest that CCHF should be far more common than it really is. However, the low rate of viraemia in large mammals, the fact that the virus is present in the blood for only a few days, and that people are not often bitten by adults of *H. m. rufipes*, *H. truncatum* and *R. e. evertsi* indicates that the disease is not as prevalent as might be expected. The immature stages of the ticks acquire the virus from their hosts *Lepus saxatilis* and *L. capensis*. The virus survives moulting to persist in unfed adult ticks and can infect humans or other large mammals if bitten by infected adult ticks.

1 Introduction

Crimean-Congo hemorrhagic fever (CCHF) virus was first isolated in 1944 when approximately 200 people sleeping outdoors in the Crimean Peninsula developed hemorrhagic symptoms. Subsequent periodic epidemics of the disease have occurred in eastern Europe, USSR, Bulgaria, Pakistan, Iraq and Egypt (Hoogstraal 1979; Tantawi et al. 1980). Later, the CCHF virus was isolated in the Congo from a child suffering from hemorrhagic fever, and in African countries such as Kenya, Ethiopia and Nigeria (Hoogstraal 1979). The virus was first isolated in South Africa during February 1981 from the blood of a 13-year-old boy who died after camping in a nature reserve (Gear et al. 1982).

Although the virus has been associated with 28 species of ixodid ticks, it is believed that the principal vectors of the virus are ticks of the genus *Hyalomma* (Hoogstraal 1979). In addition to sympatric distribution between CCHF virus and *Hyalomma* ticks, there is evidence that CCHF virus and *Hyalomma* ticks evolved together (Hoogstraal 1979). There are also indications that ticks from the genus *Rhipicephalus* may be a supplemental reservoir and should not to be disregarded as a vector (Swanepoel et al. 1983; Rechav et al. 1987). The objectives of this review are to consider the ecology of the vectors of CCHF in South Africa and to discuss the role of the various tick species in the epidemiology of CCHF.

[1] Department of Biology, Medical University of South Africa, Medunsa 0204, South Africa

2 Geographical Distribution of Vectors of CCHF

The distribution of various *Hyalomma* species in Europe, Asia, Middle East, North and Central Africa was reviewed in detail by Hoogstraal (1979). Furthermore, one can expect to find CCHF virus in the presence of *Hyalomma* ticks. These ticks are common in Eurasia and Africa and are usually found in similar habitats to that of hares or hedgehogs (Causey et al. 1970). Sufficient documentation on the distribution of *Hyalomma* species (Theiler 1956; Rechav et al. 1987) and *Rhipicephalus evertsi evertsi* (Theiler 1962) is available for South Africa. These species all prefer semi-arid habitats, while avoiding the coastal belt and higher mountain ranges of Natal and Lesotho. *H. truncatum* is, however, more sparsely distributed in the higher rainfall areas than the other species. Not only does *H. m. rufipes* have a similar distribution to that of *H. truncatum* (Theiler 1962), but their two-host life cycles are also similar.

The geographical distribution of species serving as vectors in laboratory experiments has not been included. Uncertainty as to their role in transmitting the virus under natural conditions still prevails (Hoogstraal 1979).

3 Hosts and Seasonal Activity

The seasonal activity of the three two-host tick species serving as potential vectors for CCHF virus was monitored during various surveys conducted in different parts of South Africa. The results dealing with the immature stages of *H. truncatum*, *H. m. rufipes*, and *R. e. evertsi* are presented separately to those of the adult ticks, and associated hosts.

3.1 Immature Stages

Two methods were used for monitoring the immature stages:
1. Collection of questing larval ticks from vegetation by a dragging technique (Rechav 1982, 1986; Rechav et al. 1987) – dragging a 1-m square cloth over a random path of 100 m.
2. Removal of feeding ticks from their hosts.

Free-living larvae of *H. m. rufipes* and *H. truncatum* were collected from vegetation for 36 consecutive months with two peaks of abundance in mid-winter (July) and mid-summer (November) being observed (Fig. 1). Similar peaks of abundance were observed for larvae and nymphs of the two *Hyalomma* species which were removed from two species of hares (*Lepus saxatilis* and *Lepus capensis*) over the same period (Figs. 2, 3). Hares appear to be the preferred hosts for the immature stages of *H. truncatum* and *H. m. rufipes*. However, of the two *Hyalomma* species more *H. truncatum* larvae and nymphs were removed. Similar patterns of seasonal abundance were demonstrated by larvae and nymphs of *H. truncatum* found on two species of rodents and the highveld gerbil *Tatera brantsii* and the four striped mouse *Rhabdomys pumilio*. No immature stages of *H. m. rufipes* were found on these rodents. Larvae and nymphs of *H. m. rufipes*

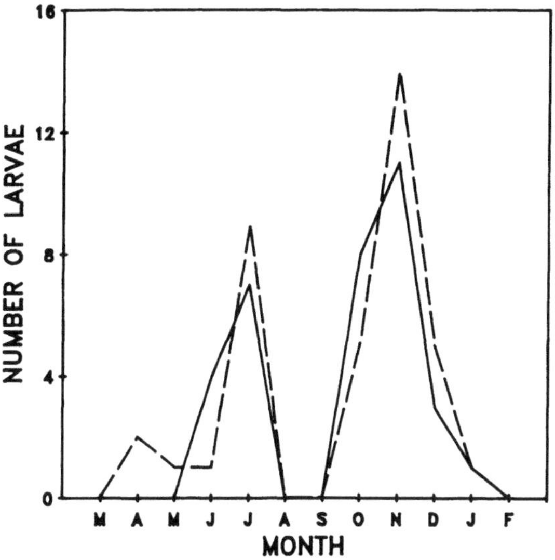

Fig. 1. Mean number of *Hyalomma truncatum* (——) and *Hyalomma marginatum rufipes* (----) larvae collected by dragging technique from a nature reserve in the western Transvaal (South Africa)

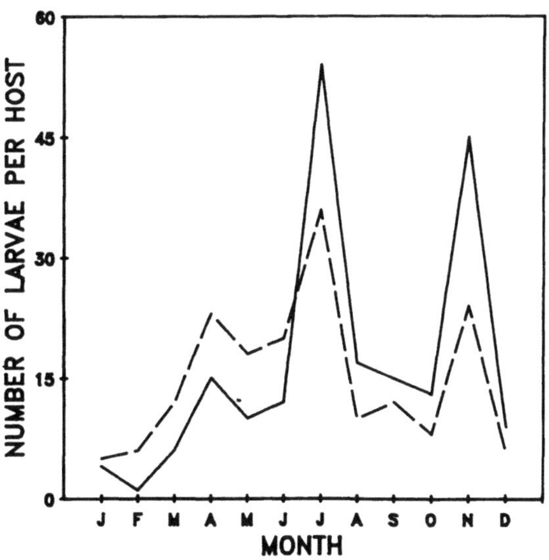

Fig. 2. Mean number of *Hyalomma truncatum* (——) and *H. marginatum rufipes* (----) larvae removed monthly from five specimens of *L. saxatilis* and *L. capensis* from a nature reserve in the western Transvaal (South Africa)

were also present on Guinea fowl, *Numida meleagris* during winter. The summer peak observed on mammal hosts was absent on Guinea fowl, probably as a result of reduced activity by Guinea fowl during nesting in October and November.

Free-living larvae of *R. e. evertsi* were collected using the drag technique. Peaks in abundance between early summer (September) and mid-summer (December) (Fig. 4) were observed. However, two peaks of abundance during

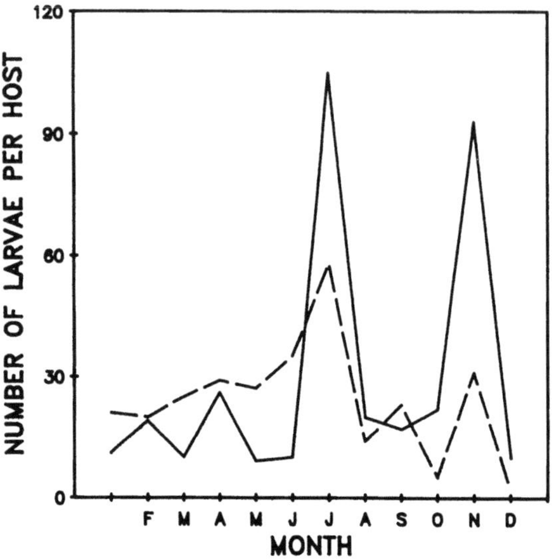

Fig. 3. Mean number of *Hyalomma truncatum* (——) and *H. marginatum rufipes* (----) nymphs removed monthly from five specimens of *L. saxatilis* and *L. capensis* from a nature reserve in the western Transvaal (South Africa)

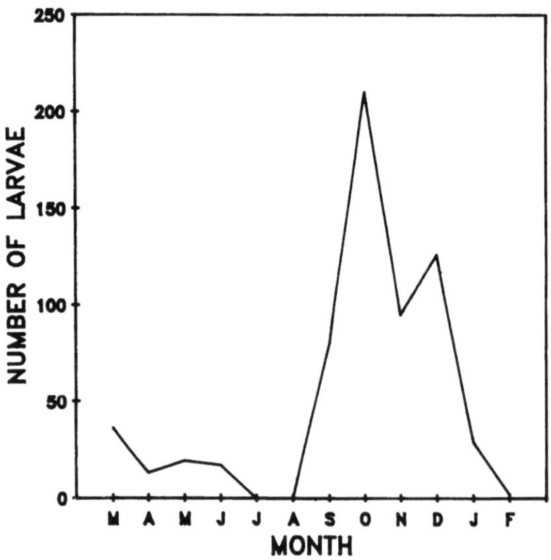

Fig. 4. Mean number of *R. e. evertsi* larvae collected by dragging technique from a nature reserve in the western Transvaal (South Africa)

autumn (March) and mid-summer (November) were observed when larvae and nymphs of *R. e. evertsi* were removed from the scrub hare *L. saxatilis* and the Cape hare *L. capensis* (Fig. 5). Immature stages of *R. e. evertsi* collected from large mammals such as impala (*Aepyceros melampus*) and red hartebeest (*Alcelaphus buselaphus*) also demonstrated two peaks of abundance in March and November.

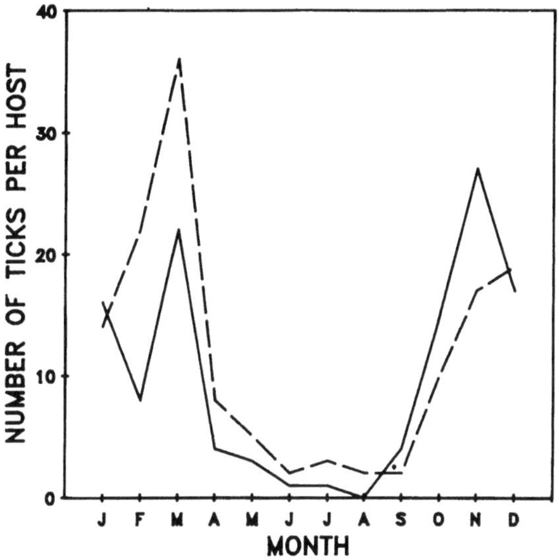

Fig. 5. Mean number of *R. e. evertsi* larvae (——) and nymphs (----) removed monthly from five specimens of *L. saxatilis* and *L. capensis* in the Lombard Nature Reserve (South Africa)

2.3 Adult Stages

Both species of *Hyalomma* demonstrated a similar pattern of seasonal abundance. Adult *Hyalomma* ticks were present on cattle during summer, being more abundant between February and Arpil (Figs. 6 and 7).

H. m. rufipes and *H. truncatum* adults collected from game animals showed a similar seasonal abundance to that found on cattle. Host-specific preference

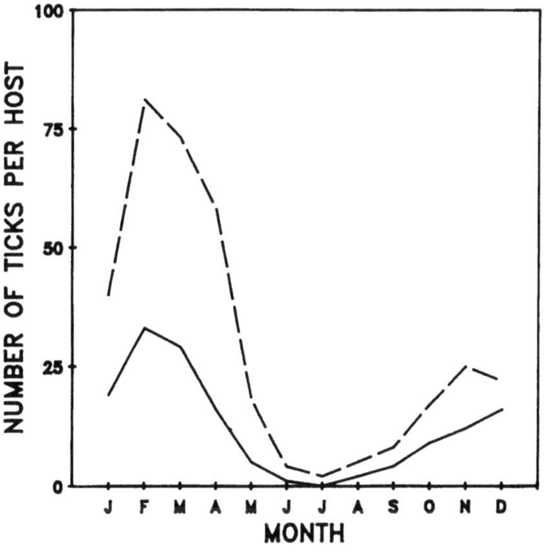

Fig. 6. Mean number of *H. truncatum* males (----) and females (——) removed monthly from eight species on three farms in the northern Transvaal (South Africa)

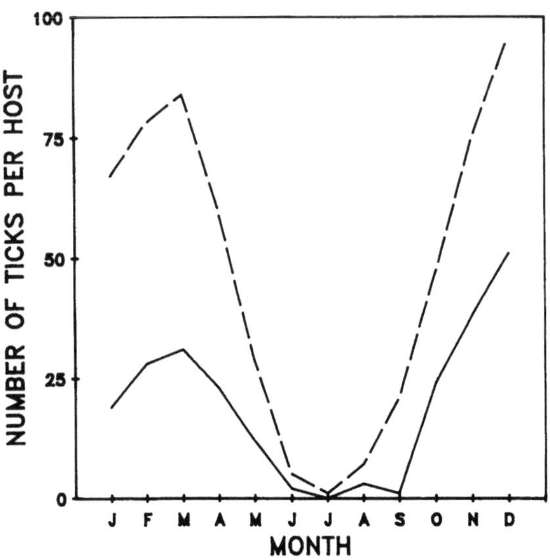

Fig. 7. Mean number of *H. m. rufipes* males (----) and females (——) removed monthly from eight species on three farms in the northern Transvaal (South Africa)

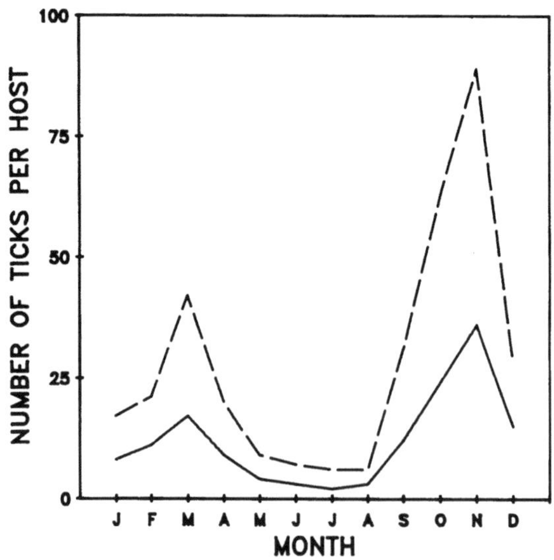

Fig. 8. Mean number of *R. e. evertsi* males (----) and females (——) removed monthly from eight species on three farms in the northern Transvaal (South Africa)

exists, resulting in some game animals such as eland, *Taurotragus oryx* and Burchell's zebra, *Equus burchelli*, carrying more *H. m. rufipes*. Other game animals such as black wildebeest, *Connochaetes gnou*, blesbok, *Damaliscus dorcas phillipsi* and red hartebeest were infested mainly with *H. truncatum* adults. *R. e. evertsi* adults were present on cattle throughout the year. However, two peaks of abundance were observed during October and February/March (early and late summer) (Fig. 8).

Adult ticks of *R. e. evertsi* were present in low numbers throughout the year on most game animals. As on cattle they demonstrated two peaks of abundance – one in the early summer and the second one during the late summer.

4 The Epidemiology of CCHF

Seasonal activities of the ticks, *H. m. rufipes*, *H. truncatum* and *R. e. evertsi*, and the infestation of various animals which serve as hosts indicate that these ticks are well established in CCHF areas (Rechav et al. 1977; Knight et al. 1978; Rechav et al. 1987). CCHF virus has been isolated from specimens of these species collected in the field. It has been shown that adults of these tick species successfully transmitted the virus to susceptible sheep (Shepherd et al. 1989a). In South Africa CCHF is transmitted by *H. truncatum*, *H. m. rufipes* and *R. e. evertsi*(Hoogstraal 1979; Swanepoel et al. 1983; Shepherd et al. 1989a). Migrating birds might carry immature stages, particularly those of two-host ticks such as *H. m. rufipes* and *R. e. evertsi*, which have a long attachment time on their hosts (Rechav et al. 1977; Shepherd et al. 1989a). Although Kaiser et al. (1974) stated that "the wide distribution of the CCHF virus is probably due to intercontinental carriage of infected ticks by migrating birds", it seems that this is at present negligible in South Africa, because the disease is already well established. The wide distribution of the CCHF virus in South Africa suggests that any introduction of the virus by ticks transported by birds from Eurasia probably happened in prehistoric times. The virus would then probably have circulated between ticks and large and small mammal hosts (Shepherd et al. 1987; Shepherd et al. 1989a) spreading to establish itself through the rest of the country. All the data indicate that adults of *R. e. evertsi*, *H. m. rufipes* and *H. truncatum* (Hoogstraal 1979; Swanepoel et al. 1983) are the main vectors of CCHF. However, experiments conducted recently showed that the viraemia of CCHF in artificially infected animals is not sufficiently high to infect feeding adults of *H. m. rufipes* (Shepherd et al. 1989b). It seems that it is probably the larvae that acquire the virus from the hosts on which they feed. Although immature stages of the two *Hyalomma* species and of *R. e. evertsi* were present on hares, rodents and other small mammals, antibodies specific to CCHF virus were found mainly in hares (40 out of 293 examined) and not in any other host collected during various surveys (Shepherd et al. 1987; 1989b). The results of this study indicate that scrub hares *L. saxatilis* developed CCHF viraemia of an intensity to be sufficient for infection of feeding immature ticks. South African hedgehogs, *Erinaceus frontalis* and wild rodents are unlikely to be of importance as hosts for the virus in southern Africa (Shepherd et al. 1989a). There is a positive correlation between the distribution of the two *Hyalomma* species and the presence of antibodies to CCHF in the blood of large mammals on which the adult ticks feed. That antibodies were not found in large mammals inhabiting areas from which hares were absent, and that no transovarial transmission of the virus occurs, supports the hypothesis that immature ticks, rather than adult ticks acquire the infection. The virus survives the moulting process to persist in adult ticks, which can then infect their hosts during feeding.

Less is known of the role of *R. e. evertsi* in the epidemiology of CCHF, firstly because this tick is found only in some parts of Africa (Theiler 1962; Rechav 1982, 1986) and secondly because the presence of CCHF virus in this tick is only a recent discovery. However, the presence of *R. e. evertsi* larvae on hares, which may lead to acquisition of viral infection, and the lack of host specificity observed in adult ticks makes *R. e. evertsi* a potential vector of the CCHF virus. More information is required before the role of *R. e. evertsi* in the epidemiology of tick-borne CCHF can be assessed.

H. m. rufipes, *H. truncatum*, and *R. e. evertsi*, the three species from which CCHF virus has been isolated, feed on various medium- and small-sized mammals as well as on ground-feeding birds. Most of these animals, except hares and probably hedgehogs, were found to be free of antibodies to CCHF (Shepherd et al. 1987). However, surveys conducted in South Africa, (Rechav 1982; Rechav et al. 1987) reports from Kenya (Clifford et al. 1976) and Nigeria (Causey et al. 1970), and the presence of immature stages of *H. truncatum* on rodents in the Lombard Nature Reserve (South Africa), provide sufficient evidence for the role of rodents, medium-sized mammals, and Guinea fowl in supporting populations of tick species known to be important to the epidemiology of CCHF. Large mammals undoubtedly serve as amplifiers for populations of *H. m. rufipes*, and *H. truncatum*, and *R. e. evertsi*. The large number of ticks and the presence of many hosts allow successful transcription between ticks and mammals and the establishment of the virus in South Africa.

Acknowledgements. I am most grateful to the Transvaal Nature Conservation Division and to L.G. Oats, who kindly allowed the use of the S.A. Lombard Nature Reserve; and to the late Harry Hoogstraal (Naval Medical Research Unit No. 3) for fruitful discussions, and to Dr L.J. Fielden and A.J. Mackie from the Department of Biology, Medunsa, for revising the manuscript.

References

Causey OR, Kemp GE, Madboudly MH, David-West TS (1970) Congo virus from domestic livestock, African hedgehog, and arthropods in Nigeria. Am J Trop Med Hyg 19:846–850

Clifford CM, Flux JEC, Hoogstraal H (1976) Seasonal and regional abundance of ticks (*Ixodidae*) on hares (*Leporidae*) in Kenya. J Med Entomol 13:40–47

Gear JHS, Thompson PD, Mopp M, Andronikye S, Cohn RJ, Ledger J, Berkowitz FE (1982) Congo Crimean haemorrhagic fever in South Africa. Report of a fatal case in the Transvaal. S Afr Med J 62:576–580

Hoogstraal H (1979) The epidemiology of tick-borne Crimean-Congo haemorrhagic fever in Asia, Europe and Africa. J Med Entomol 15:307–417

Kaiser MN, Hoogstraal H, Watson GE (1974) Ticks (*Ixodidae*) on migrating birds in Cyprus, fall 1967 and spring 1968, and epidemiological considerations. Bull Entomol Res 64:97–100

Knight MM, Norval RAI, Rechav Y (1978) The life cycle of the tick *Hyalomma marginatum rufipes* Koch (Acarina: Ixodidae) under laboratory conditions. J Parasitol 64:143–146

Rechav Y (1982) Dynamics of tick populations (*Acari*: *Ixodidae*) in the Eastern Cape province of South Africa. J Med Entomol 19(6):679–700

Rechav Y (1986) Seasonal activity and hosts of the vectors of Crimean-Congo haemorrhagic fever in South Africa. S Afr Med J 69:364–368

Rechav Y, Knight MM, Norval RAI (1977) The life cycle of the tick *Rhipicephalus evertsi evertsi*. J Parasitol 63:575–579

Rechav Y, Zeederberg ME, Zeller DA (1987) Dynamics of African tick populations in a natural Crimean-Congo hemorrhagic fever focus. J Med Entomol 24:575–583

Shepherd AJ, Swanepoel R, Shepherd SP, McGillivary GM, Searle LA (1987) Antibody to Crimean-Congo haemorrhagic fever virus in wild mammals from southern Africa. Am J Trop Med Hyg 36:133–142

Shepherd AJ, Swanepoel R, Cornel AJ, Mathee O (1989a) Experimental studies on the replication and transmission of Crimean-Congo haemorrhagic fever virus in some African tick species. Am J Trop Med Hyg 40:326–336

Shepherd AJ, Leman PA, Swanepoel R (1989b) Viremia and antibody response of small African and laboratory animals to Crimean-Congo haemorrhagic fever virus infection. Am J Trop Med Hyg 40:541–547

Swanepoel R, Struthers JK, Shepherd AJ, McGillivray GM, Nel MJ, Jupp PG (1983) Crimean-Congo haemorrhagic fever in South Africa. Am J Trop Med Hyg 32:1407–1415

Tantawi MK, Al-Moslin MI, Hassan FK, Al-Ani FS (1980) Crimean-Congo haemorrhagic fever. N 3, T–1474

Theiler G (1956) Zoological survey of the Union of South Africa: tick survey: Part ix. The distribution of the three South African Hyalommas or bontpoot Onderstepoort J Vet Res 27:239–269

Theiler G (1962) The Ixodidae parasites of vertebrates in Africa, south of the Sahara (Ethiopian Region). Report to the Director of Veterinary Services, Onderstepoort

Protection of Cattle Against Babesiosis in Tropical and Subtropical Countries with a Live, Frozen Vaccine

A.J. de Vos[1] and W.K. Jorgensen[1]

Summary

Babesiosis in Africa is usually caused by *Babesia bovis* and/or *Babesia bigemina*. Their distribution follows the tick vectors *Boophilus microplus* (*B. bovis* & *B. bigemina*), *Boophilus geigyi* (*B. bovis* & *B. bigemina*), *Boophilus decoloratus* (*B. bigemina*) and *Rhipicephalus evertsi evertsi* (*B. bigemina*). The prevalence of clinical disease in a locality depends on transmission rates. Enzootically stable areas have high transmission rates and animals are infected during the first 8 months of life. Up to this age they are protected by maternal antibodies and non-specific age-related factors, and infection does not usually result in clinical disease but induces protective immunity. Enzootic instability is caused by low transmission rates that may be due to environmental factors or management practices such as dipping, resulting in a proportion of the adult cattle population remaining susceptible to the disease. Severe disease outbreaks can occur amongst susceptible cattle when vector numbers and thus transmission rates increase. Similarly, high mortality can occur when susceptible cattle are imported to localities where the disease is endemic.

The most effective protection currently available for susceptible cattle is vaccination with a living vaccine from which cattle acquire a long-lasting immunity. Chilled, live babesiosis vaccine is produced in Australia and South Africa. The vaccine has a short shelf life making distribution and post-production testing for efficacy and contamination impractical in some countries. The requirements of a longer shelf life and quality control after production have led to development of a new Australian babesiosis vaccine cryopreserved in liquid nitrogen.

Choice of *Babesia* vaccine strains and specific requirements for vaccine production are discussed in relation to problems that may be encountered in Africa when establishing a vaccine production facility. Application of in vitro *Babesia* culture technology to vaccine production is also discussed.

Since the introduction of a standardized method of vaccine production in Australia, living babesiosis vaccines have generally proved highly effective. Clinical disease outbreaks have, however, occurred in vaccinated herds and possible reasons are suggested. Future work will investigate the mechanisms of antigenic variation in *Babesia* species in relation to host selection pressure.

1 Introduction

Babesiosis, also known as redwater or tick fever, is a tick-borne disease of cattle caused by pathogenic species of the protozoan parasite *Babesia*. In many tropical and subtropical countries, disease caused by *Babesia bigemina* and *Babesia bovis* has had a marked effect on the development of livestock industries over many decades. Exotic *Bos taurus* breeds of cattle brought to these countries to improve the production potential of livestock are particularly susceptible, sometimes with devastating results following exposure to the combined effects of

[1] Tick Fever Research Centre, Department of Primary Industries, Wacol, Queensland, Australia

babesiosis, other tick-borne diseases and tick infestations per se (Callow 1978; Lawrence and Norval 1979; McCosker 1981; Pipano and Hadani 1984).

Attempts to control bovine babesiosis have included various methods ranging from vaccination to vector control. Early attempts to immunize cattle involved use of crude vaccines produced from blood of recovered cattle, a technique pioneered in Australia at the turn of the century (Callow and Dalgliesh 1980). With some significant improvements in the method of production, use of live vaccine prepared from infected blood, is still the most effective method for immunizing cattle (Pipano and Hadani 1984; de Vos et al. 1987).

In this paper, we discuss bovine babesiosis caused by *B. bigemina* and *B. bovis*, particularly in the relation to the occurrence of the disease in Africa. We also discuss the technique used to produce highly effective, frozen, live babesiosis vaccine in Australia and its possible application in African countries.

2 Causative Organisms

Babesia bovis, the most pathogenic of the bovine *Babesia* spp., is transmitted in Africa mainly by *Boophilus microplus*, a tick introduced during the latter part of the previous century (Potgieter 1977; Norval and Short 1984). This tick is most prevalent in high rainfall areas in the south eastern part of the continent and along the Mediterranean. Another tick species, *Boophilus geigyi*, is a vector of *B. bovis* in West Africa (Friedhoff 1987).

Babesia bigemina is transmitted in Africa by the common cattle tick, *Boophilus decoloratus*, as well as by *B. microplus*, *B. geigyi* and *Rhipicephalus evertsi evertsi* (Potgieter 1977; Friedhoff 1987). As a result of the prevalence and wide distribution of these vectors, *B. bigemina* is more widespread than *B. bovis* (de Vos 1979; Norval et al. 1983) and, for this reason, is often considered to be the more important of the two species, even though it is less pathogenic than *B. bovis*.

Babesia occultans is a little known parasite of cattle in Africa transmitted by *Hyalomma marginatum rufipes* (Gray and de Vos 1981). It is relatively benign and therefore of little or no economic significance.

3 Epidemiology and Indications for Control

3.1 Enzootic Stability

The occurrence of babesiosis in a cattle population depends largely on the frequency with which the causative species is transmitted. If there is frequent transmission of the parasite, all calves will become infected during the first 6 to 8 months of life. Calves in this age group are protected by maternally acquired and non-specific factors and immunity will develop without evidence of disease (Mahoney and Ross 1972; Callow 1977). This situation is known as enzootic stability.

Millions of cattle in Africa and other tropical and subtropical regions survive because of the presence of a state of enzootic stability and the genetic resistance of the predominantly indigenous cattle involved (Pipano and Hadani 1984; Lawrence et al. 1991).

3.2 Enzootic Instability

Instability and hence the risk of disease increases when the rate of transmission decreases to levels where some animals in the population fail to become infected early in life (Callow 1977). Management and environmental factors contribute to the development of this state of instability and, in Africa in particular, the deliberate reduction of tick numbers through dipping plays a major role. Frequent dipping of cattle is often required to limit the effects of harmful multi-host ticks such as *Rhipicephalus* spp., to reduce the risk of transmission of diseases such as East Coast fever and heartwater or to comply with government regulations (Lawrence and Norval 1979; Norval 1983). The single host *Boophilus* spp. spend up to 3 weeks on a host and are therefore vulnerable under these conditions. Reductions in the numbers of these ticks are likely to be reflected ultimately in increased levels of enzootic instability.

Much of Africa is highly favourable for *Boophilus* spp. (Sutherst and Maywald 1987) although environmental factors such as rainfall, droughts and temperature (altitude) render parts of the continent at times very marginal for survival of these ticks. Furthermore, differing requirements and interspecific competition between *B. microplus* and *B. decoloratus* (Norval and Short 1984) result in a zone of interference between the habitats of the two species. This zone has no effect on the transmission rate of *B. bigemina*, a parasite transmitted by both tick species, but may have a very significant effect on the transmission of *B. bovis*.

A simple mathematical model has been used to predict the level of stability in a herd (Mahoney and Ross 1972). This model utilizes the rate at which infection occurs in calves as measured by serological means, and has been used to estimate the enzootic status of *B. bovis* infections in *Bos taurus* cattle. It has also been used to predict the level of stability for *B. bigemina* in South Africa (de Vos and Potgieter 1983). However, the model's predictions may not be accurate for Sanga, Zebu and cross-bred cattle due to the low numbers of ticks infesting these animals (Mahoney et al. 1981).

One tick is sufficient to transmit the infection (Mahoney and Ross 1972) but not all ticks are infected. In an Australian field study, only 0.04% of larval ticks were found to be infected with *B. bovis* when *Bos taurus* cattle were involved and even less in the case of zebu cattle (Mahoney et al. 1979, 1981). Tick infection rates for *B. bigemina* are considerably higher (0.23% in the Australian study) and transmission rates for this species are therefore also higher than those of *B. bovis*. As a result, *B. bigemina* is the more prevalent of the two species in herds where both are present (de Vos 1979) and therefore less likely to be affected by managerial and environmental factors which reduce tick numbers.

Use of a babesiosis vaccine is indicated where enzootic instability is an ongoing cause of losses and where the factors contributing to this situation are likely to remain for an indefinite period.

3.3 Epizootic Spread

The distribution of *B. bigemina* and *B. bovis* is limited by environmental factors such as rainfall and also by tick control measures. Removal of these barriers is likely to result in the tick vectors becoming established temporarily or permanently in previously uninfested areas. This happened when tick control measures broke down during a period of political unrest in Zimbabwe and *B. microplus* (and with it *B. bovis*) spread from a few foci near the Mozambique border to the centre of the country (Norval et al. 1983). A period of unusually high rainfall also caused *B. decoloratus* and *B. bigemina* to temporarily spread several hundred kilometres through a normally arid part of South Africa (de Vos 1979).

Restricted movement of tick-infested cattle and introduction or reintroduction of effective tick eradication measures are logical steps which will assist in restoring the pre-epizootic status quo. However, in naturally tick-endemic regions, the cost of such measures should be weighed against those of alternative options such as vaccination against babesiosis (Norval 1983). The long-term suitability of the environment for tick survival and the need to control vectors of other tick-borne diseases will obviously have a major bearing on the strategy to be adopted.

3.4 Cattle Imported into an Endemic Area

Babesiosis is often the cause of severe losses in susceptible cattle, especially when *Bos taurus* cattle are introduced into *Boophilus*-infested parts of Africa. Mortality rates of 5 to 10% under these conditions are quite common (de Vos 1979). Importation of totally naive cattle as in the case of foreign aid programmes, is known to have resulted in mortality rates exceeding 80% due to ticks and their transmitted diseases (including babesiosis) (A.J. de Vos, unpubl. observ. 1990). As imported cattle are usually very valuable, efficient tick-borne disease control programmes are key elements in the design of genetic improvement projects in tick-infested regions of Africa and elsewhere. Use of an effective babesiosis vaccine must be an integral part of these projects.

4 Control of Babesiosis by Vaccination

4.1 General Principles

The majority of effective immunization strategies for control of bovine babesiosis are based on the use of living organisms as vaccine. The purpose of using these vaccines is to give cattle a mild form of the disease which induces lasting protection against subsequent natural infections. Most strains used in vaccine consist of a variety of heterogeneous, separable subpopulations (Carson et al. 1990; Timms et al. 1990).

Based on experimental evidence it is known that immunity can also be induced with crude or purified antigens prepared from infected blood (Mahoney et al. 1984). However, progress in the development of non-living vaccines based on these antigens has been slow because of the complex nature and intra-erythocytic development of the organisms, and difficulties in obtaining adequate amounts of purified antigen (Mahoney et al. 1984). Initial attempts to isolate suitable antigens were aimed mainly at infected erythrocytes but, more recently, exoantigens derived from supernatant of *B. bovis* cultures have also been shown to have immunogenic properties (Timms et al. 1983; Montenegro-James et al. 1989). However, the level and duration of protection conferred by these exoantigens against challenge with heterologous strains has been inferior to that of live vaccine (Timms et al. 1983). To solve the dual problem of production and purification of protective antigens, the aim will be to clone and express antigens in suitable vectors (Mahoney et al. 1984). Commercial viability of further genetically engineered vaccines will depend on the level of protection afforded against virulent field infections and on the potential effect of antigenic change in field strains. As reviewed by de Vos et al. (1987), the ability of field strains of *Babesia* to change antigenically is well known, but the effect of these changes on the long-term efficacy of vaccines has received little attention (Walliker 1989). Most *B. bovis* strains isolated in Australia have very few common alleles which indicates rapid genomic change (Dalrymple and Wright 1991, unpubl. observ.)[2]. Unfortunately, the ability of these parasites to change does not bode well for the future of genetically engineered vaccines containing a small number of expressed antigens.

4.2 Carrier-Donor Vaccine

Carrier-donor vaccine is produced from blood of artificially infected cattle, usually after clinical recovery of the animals (Callow 1977; Anonymous 1984). The vaccine was popular in the past as it is simple to produce and can be prepared on private properties (Lawrence and Norval 1979). Unfortunately, variable infectivity of blood from carrier animals has resulted in unacceptably high failure rates for this vaccine and this method of production has fallen into disrepute in most countries.

---

[2] B.P. Dalrymple and I.G. Wright, CSIRO Longpocket Laboratories, Meiers Road, Indooroopilly, Queensland, Australia

4.3 Standardized Chilled Vaccines

The variable infectivity of the blood of recovered animals led to the development in Australia of a highly infective chilled vaccine prepared from blood of acutely infected, splenectomized calves (Callow 1977; Callow and Dalgliesh 1980). The vaccine contains 1×10^7 parasites per dose, a number known from infectivity trials to be about 100 times the minimum infective dose (Callow 1977). Heavily infected blood is collected in an anticoagulant, and the parasites quantified and diluted to the required number in a sustaining solution containing serum, glucose and balanced salts (Callow 1977). If kept chilled, this vaccine will remain viable for a week from the time of production.

Chilled babesiosis vaccine enjoys great popularity in Australia, largely because of its low cost and proven efficacy (Callow and Dalgliesh 1980). It is also used with good results in South Africa (de Vos 1979). However, there are constraints in the production and distribution of this vaccine which cause concern and will put production beyond the practical reach of some countries.

Firstly, the perishable nature of chilled vaccine makes it impossible to evaluate product quality before dispatch. Every effort must therefore be made to ensure the donors are free of harmful pathogens at the time of use. Depending on the pathogens involved, these efforts may have to be intense and expensive. In Australia, exhaustive steps are taken to select, screen and house donor calves but, despite these precautions, one batch of vaccine produced during 1986 was later shown to have been contaminated with bovine leucosis virus (BLV) (Rogers et al. 1988).

Secondly, chilled vaccine requires an integrated, reliable transport network for rapid conveyance after production, preferably with delivery within 24 h.

Thirdly, the short shelf life makes it impossible to utilize more than a fraction of the full potential of each donor (about 50 000 doses).

4.4 Standardized Frozen Vaccine

The need for proper quality control of babesiosis vaccine has led to the development of frozen vaccine (Mellors et al. 1982; Pipano and Hadani 1984; Dalgliesh et al. 1990). This vaccine allows thorough testing for both efficacy and the absence of contaminants before release.

Frozen vaccine shares the benefit of the chilled product. It is highly effective and induces lasting protection after a single inoculation. In addition, long life in the frozen form (5 years in liquid nitrogen) makes it possible to fully exploit each donor used. The advantages of the frozen babesiosis vaccine makes it the product of choice but the availability of disease-free donors and storage facilities (liquid nitrogen, dry ice or − 70 °C freezers) will obviously be major constraints in some countries. Transportation of the frozen product is also likely to present difficulties in some instances.

4.5 Choice of Strains for Production of Vaccine

There are two options when selecting strains for production of vaccine. The first is to import a strain that is known to be effective, of low virulence and free from contaminants. Australian strains of *B. bovis* and *B. bigemina* have been shown experimentally to be protective in South Africa (de Vos et al. 1982a, b); Sri Lanka (W.K. Jorgensen, unpubl. observ. 1986) and Bolivia (Callow et al. 1976). Vaccine containing these strains has also been used with beneficial results in countries in many parts of the world, including Malawi and Swaziland in Africa, Venezuela and Equador in South America, Malaysia and Thailand in South-East Asia and islands in the Caribbean.

A second, often tempting option is to isolate strains from the country in which the vaccine is to be used, to attenuate these strains and to remove contaminants. Strains can be isolated by feeding infected ticks on a susceptible bovine or by inoculating an animal with blood from a carrier (Anonymous 1984). *B. bigemina* parasites with low virulence can be obtained by relatively "slow" passage of a strain in intact calves and then splenectomizing the calves 3 months after the primary parasitemia (Dalgliesh et al. 1981; Anonymous 1984). Although this is not assured, parasites obtained during the post-splenectomy relapse are often of low virulence and therefore suitable for production of vaccine. Rapid calf-calf passage of *B. bigemina* is not recommended as it may result in an increase in the virulence of the organisms. *B. bovis*, in contrast, can usually be attenuated by a series of rapid passages through up to 20 splenectomized calves. This procedure exerts selective pressures on parasite populations which result in survival of predominantly avirulent organisms (Callow 1977; Anonymous 1984). Irradiation (Wright et al. 1980) and long-term maintenance of *B. bovis* in culture containing horse serum (Yunker et al. 1987) have also been reported to have attenuating effects but, again, this result is not assured (Weilgama et al. 1989; W.K. Jorgensen, unpubl. observ. 1990). A major impediment to the use of local strains in vaccine is the presence of potentially harmful contaminants. In Australia, *B. bovis* and *B. bigemina* strains have been maintained for one month or more in vitro as described by Levy and Ristic (1980) and Vega et al. (1985) to remove contaminants which do not multiply under these conditions such as *Theileria*, *Anaplasma*, *Trypanosoma* and *Eperythrozoon* spp. (W.K. Jorgensen, unpubl. observ. 1990). However, absence of adventitious viral infections can only be ensured if the donor cattle are obtained from, or bred in disease-free areas and maintained under tick- and insect-free isolation conditions.

Suitable, contaminant-free strains should be cryopreserved for use when required.

5 Production of Frozen Babesiosis Vaccine in Australia

The following procedures are employed in Australia to produce safe, effective frozen babesiosis vaccine for local use and for export.

5.1 Preparation of Cryopreserved Stabilates

Attenuated contaminant-free strains are stored as cryopreserved stabilates in liquid nitrogen and can be kept in this way for many years without significant changes in virulence or immunogenicity. After thawing, the stabilates are used to infect donor cattle for preparation of vaccine.

Stabilates are prepared with dimethyl sulphoxide (DMSO) as cryoprotectant (Dalgliesh 1972; Mellors et al. 1982). Briefly, parasitized blood is collected in heparin, ACD or CPD and chilled to 4 °C. The blood is then slowly mixed with an equal volume of 4 M DMSO in PBS at 4 °C, aliquoted in 5 ml cryovials and frozen in the gas phase of liquid nitrogen at a cooling rate of about 10 °C/min. When frozen, the vials are transferred into liquid nitrogen for long-term storage.

5.2 Source of Parasites for Preparation of Vaccine

A vial of stabilate is thawed rapidly by immersion in water pre-heated to 37–40 °C. When thawed, it is placed on ice and the contents used as soon as possible to inoculate a donor animal (within 20–30 min). Stabilate prepared with DMSO can be inoculated by the intravenous route to reduce the time to the expected reaction by several days compared with that following other routes of inoculation.

Parasites for production of frozen vaccine can be obtained in the same way as for chilled vaccine. Traditionally, this has been done by inoculating a susceptible, splenectomized calf with infective inoculum. The calf is then monitored by examination of stained blood films, parasite numbers quantified and blood collected for vaccine when a suitable parasitemia is reached (Callow 1977; Anonymous 1984). Parasite counts of at least 1×10^8/ml of blood are satisfactory for production of frozen *B. bovis* vaccine while as little as 3×10^7 *B. bigemina* parasites/ml have been used successfully (Jorgensen et al. 1989a).

5.3 Preparation of Vaccine

Frozen vaccine is prepared in much the same way as frozen stabilate. DMSO has been used by various workers to prepare the vaccine (Mellors et al. 1982; Pipano and Hadani 1984) but this product has the disadvantage of having to be used within 20–30 min after thawing. As a result, commercial production of frozen vaccine in Australia was recently converted to using glycerol as cryoprotectant (Jorgensen et al. 1989a; Dalgliesh et al. 1990). The main advantage of glycerol over DMSO is the relatively long viability of the vaccine after thawing (at least 8 h).

The method for producing frozen babesiosis vaccine with glycerol has been discussed by Jorgensen et al. (1989a) and Dalgliesh et al. (1990). Briefly, infected blood is slowly mixed at 37 °C with an equal volume of pre-heated 3 M glycerol in PBS supplemented with 5 mM glucose (final concentration of glycerol 1.5 M).

The mixture is then incubated at 37 °C for 30 min with occasional mixing to allow for equilibration and dispensed in 5 ml cryovials suitable for storage in liquid nitrogen.

The filled vials are frozen in the gas phase of liquid nitrogen (cooling rate 10 °C/min) and stored in the liquid phase until required.

5.4 Quality Control

The quality control of each batch of frozen babesiosis vaccine is aimed at determining its effectivity (infectivity) and confirming the absence of contaminants.

Infectivity is tested by inoculating groups of cattle and then monitoring the reactions. Dalgliesh et al. (1990) described a method for testing infectivity whereby vaccine is diluted 1 : 5 in an isotonic diluent containing 1.5 m glycerol and incubated at 30 °C for 8 h before use. Subsequently, the sensitivity of this testing procedure was improved by increasing the dilution factor to 1 : 50. Each batch is tested following dilution and incubation in groups of five cattle. The cattle are monitored for the presence of infection by examination of stained blood smears as well as retrospective serology and only batches shown to be infective are cleared for use.

Control of Contamination. A major disadvantage of any living vaccine grown in an animal host is the possibility of contamination. The level of precautions that has to be taken to prevent this from happening depends on the nature of the potential contaminants in the country involved. Due to strict quarantine procedures, Australia is in a fortunate position where many important diseases as well as pathogens are absent, including foot and mouth disease, rinderpest, lumpy skin disease, Rift Valley fever, rabies, contagious bovine pleuropneumonia, heartwater, Jembrana disease as well as pathogenic *Theileria* spp. and *Trypanosoma* spp.

The following precautions are taken to prevent or detect contamination of vaccine donors with other adventitious agents. This is done before the vaccine is produced.

1. Field infections of *Babesia* and *Anaplasma*: the donor cattle are procured from or bred in tick-free surroundings and tested for the absence of antibodies to *B. bovis*, *B. bigemina* and *A. marginale*. Thereafter, the cattle are accommodated in a tick- and insect-free environment, splenectomized and monitored for parasitic relapses by microscopic examination of stained blood smears.
2. Field infections of *Theileria buffeli*: this infection is found in 50% of all cattle used for vaccine production. Chemical sterilization of the infection is achieved with primaquine and buparvaquone (Stewart et al. 1990).
3. Enzootic bovine leucosis virus (BLV): the cattle are obtained from closed herds tested free of the infection. In addition, the cattle are tested by

inoculation of lymphocytes into sheep and then monitoring the sheep over a period of 14 weeks for evidence of seroconversion by an agar gel immunodiffusion test (Rogers et al. 1988). The sheep transmission test is the most sensitive test presently available for detection of this virus.
4. Brucellosis: cattle are obtained from properties tested free of this disease.
5. Tuberculosis: cattle are obtained from a tuberculosis-free area of Australia.
6. Ephemeral fever: the cattle are housed under insect-free conditions for at least 7 weeks prior to use.
7. Akabane disease: as for ephemeral fever.
8. Mucosal disease (MD) or pestivirus: many cattle in Australia become naturally exposed to this virus early in life but do not retain the infection for a significant period of time. However, calves born to dams naturally exposed for the first time during pregnancy may become immunotolerant and persistently viraemic. These calves pose a great risk to the production of vaccine. All donor calves are monitored for the presence of infection by the sheep transmission test concurrently with the test for BLV. An antigen capture ELISA is also being evaluated for use in conjunction with the transmission test.
9. *Eperythrozoon* spp: only cattle shown to be free from the infection by microscopic examination of blood smears are used. The cattle are housed under tick- and insect-free conditions.
10. *Leptospirosis*: all cattle are treated twice upon arrival with streptomycin at an interval of 48 hours.

Each batch of blood used for production of frozen vaccine is also tested for freedom of contaminants. This is done by inoculating a sheep with 10 ml of blood and then monitoring it serologically for the development of antibodies to BLV, MD, blue tongue, akabane virus, ephemeral fever and aino virus. A batch showing any evidence of contamination is discarded.

5.5 Use of Frozen Vaccine

B. bovis and *B. bigemina* vaccines are prepared separately in Australia using colour-coded cryovials. After thawing, each type of vaccine can be used individually or mixed to prepare bivalent babesiosis vaccine. In most cases, *Anaplasma centrale* vaccine which is prepared in exactly the same way, is included to prepare trivalent "tick fever" vaccine. Equal numbers of vials of each species are thawed rapidly by agitation in water pre-heated to 37–40 °C and the contents pooled. Prepared vaccine is kept cool and used within 8 h.

One 5 ml cryovial contains sufficient vaccine to inoculate each of seven head of cattle with a 0.7 ml dose. When thawed and mixed as trivalent vaccine, the dose per animal is 2–2.1 ml.

Inoculations can be given by the subcutaneous or intramuscular routes. Vaccine reactions are seen infrequently but may develop 5–10 days after inoculation in the case of *B. bigemina* and 8–20 days if *B. bovis* vaccine is used.

6 Factors Critical to Production of Frozen Babesiosis Vaccine in Africa

The following factors are considered critical to the production of safe, effective babesiosis vaccine (Dalgliesh 1985; de Vos and Potgieter 1991).

1. Availability of attenuated, immunogenic strains of both *B. bovis* and *B. bigemina*. Several *B. bovis* strains are known to meet these criteria (de Vos and Potgieter 1991) but only one strain of *B. bigemina*, an Australian strain referred to as "G" strain, has been adequately documented in the literature (Dalgliesh et al. 1981). Fortunately, antigenic differences which exist between strains in each species do not appear to affect the immunogenic properties of these strains, even if used in other countries (Callow et al. 1976; de Vos et al. 1982a, b; Dalgliesh et al. 1990).
2. Supply of susceptible donors. Production of vaccine relies heavily on a source of donor cattle free from tick-borne diseases and potential blood-borne viral, bacterial and rickettsial pathogens. In many African countries, this can only be achieved by breeding the donors under virtual disease-free conditions.
3. Facilities for maintaining donor cattle. Isolation and surgical facilities are necessary in order to keep donors under quarantine conditions, to perform splenectomies and to collect blood for production of vaccine.
4. Monitoring of donors. Laboratory facilities and expertise are necessary to monitor donors by serological and microscopical means for susceptibility to *Babesia* spp. Expertise or at least access to expert advice, is also essential for screening of the donors for absence of pathogens.
5. Laboratory facilities with reliable electricity supplies at constant voltage to produce and store the vaccine.
6. Access to reliable liquid nitrogen or dry ice supplies for storage and transport of vaccine. Alternatively, $-70\,^{\circ}C$ freezers can be used to store the vaccine.
7. Access to transport networks which will allow the vaccine to reach its destination with minimum delay.
8. Access to foreign capital for purchase of essential equipment for vaccine production, storage and transportation.
9. A core of trained, dedicated staff to produce the vaccine.

7 New Developments in Production of Live Vaccines

7.1 Culture-Derived Vaccine

Recent studies in Australia (Jorgensen et al. 1989b; Timms and Stewart 1989) indicate that both *B. bovis* and *B. bigemina* can be grown in adequate numbers in culture to allow production of vaccine without the need to infect donor cattle. For reasons of quality control and animal ethics, this is a much preferred method of producing parasites for vaccine. Indications are that parasites harvested from culture do not differ from those obtained from donor cattle with regard to either virulence or immunogenicity. However, observations on the genetic profiles of parasites show evidence of drifting population ratios in parasites maintained in culture for long periods (W.K. Jorgensen and B.P. Dalrymple, unpubl. observ. 1990). Until more is known of this drift, cultures should not be maintained for longer than 4 months if the parasites are to be used for preparation of vaccine.

To produce parasites for vaccine in vitro, the first step is to adapt the strain to maintenance in stationary phase cultures using the well documented techniques of Levy and Ristic (1980) and Vega et al. (1985). Stationary phase culture material is then used to initiate spinner flask (Timms and Stewart 1989) or suspension flask cultures (W.K. Jorgensen, unpubl. observ. 1989). The results of Timms and Stewart (1989) show that up to 1000 doses of vaccine can be produced every 48 h from 500 ml of culture medium. Since 1990, *B. bigemina* vaccine has been produced in Australia from culture material in a trial aimed at evaluating the viability and cost-effectiveness of this method of production (W.K. Jorgensen, unpubl. observ. 1990). The aim is to produce 60000 doses during the first year. Pending the outcome of the trial, the project will either be abandoned or expanded to include *B. bovis*.

7.2 Vaccine Failures and Antigenic Variation

Since the introduction of a standardised method of production in Australia, living babesiosis vaccine has generally proved to be highly effective (Callow and Dalgliesh 1980). In most cases, a single vaccination provided lasting, probably life-long, immunity against field infections with antigenically different strains. However, troublesome failures of *B. bovis* vaccine occurred in 1968, 1976 and again in 1989–1990. In the first two instances, this was thought to be due to loss of immunogenicity of the vaccine strains as a result of numerous syringe passages in splenectomized calves (Callow and Dalgliesh 1980). In both cases, the problem was solved by replacing the vaccine strain with a new, attenuated one.

"K" strain of *B. bovis* was included in the vaccine in 1977 and, in an attempt to prevent future recurrences of the problem, the number of passages of the strain were limited. After about 20 passages frozen stabilates were stored in liquid nitrogen and used to initiate infections in calves which were then syringe passaged for a further 10 passages at most (Callow and Dalgliesh 1980).

Since the introduction of "K" strain and this method of production, the incidence of reported vaccine failures remained at an acceptably low level until 1988 (Table 1). However, between 1989 and 1990 there has again been a progressive increase in the number of failures (Bock et al. 1992). These vaccine failures were seen only in *Bos taurus* cattle and often under conditions of good management where the cattle had been vaccinated more than once and were seropositive at the time of the outbreak. The occurrence of these failures did not correlate with time after vaccination. Instead there appeared to be a link with stock movements and there was evidence of regional spread from specific foci. Four strains isolated from affected properties since 1989 have been shown

Table 1. *B. bovis* vaccine failures related to vaccine supply 1973–1990 (Bock et al. 1992)

Year	Strain in vaccine[a]	No. of vaccine failures	No. of vaccine orders	Vaccine failures per 1000 orders
1973	A and B	17	9248	1.84
1974	A and B	15	9542	1.99
1975	A and B	10	5477	1.83
1976	A and B	20	5385	3.76
1977	C and K	6	4510	1.33
1978	C and K	6	4622	1.30
1979	C and K	7	5893	1.19
1980	K	3	6323	0.47
1981	K	6	6383	0.94
1982	K	6	6190	0.97
1983	K	5	7391	0.68
1984	K	7	7017	1.00
1985	K	5	8046	0.62
1986	K	8	7459	1.07
1987	K	10	7448	1.34
1988	K	16	8586	1.86
1989	K	24	9558	2.51
1990[b]	K	37	6848	5.40

[a] Where two strains are shown, each strain was used for a period of 6 months.
[b] In September 1990 K strain was replaced by T strain in the vaccine. Therefore the figures for 1990 are only up to the end of August.

experimentally to break through the immunity of cattle vaccinated with "K" strain.

The problem was solved in September 1990 by replacing the "K" strain vaccine with a new attenuated strain ("T" strain) known from laboratory trials to be protective against the breakthrough strains. However, selection, attenuation and testing of the new strain took 2 years during which time vaccine failure resulted in significant losses. One way to prevent this from happening again may be to introduce new strains into the vaccine production programme at regular intervals or possibly to rotate a number of strains in the programme. However, the latest spate of vaccine failures clearly showed that the underlying causes are not fully understood and require further investigation if the problem is to be solved on a permanent basis. There appear to be two possible reasons for the recent failures:

1. Loss of immunogenicity of "K" strain as apparently happened with previous strains during 1968 and 1976 (Callow and Dalgliesh 1980). This is considered unlikely as procedures were in place (use of stabilate banks, restricted passaging) to prevent gradual change of the strain. There is no assurance that infective material stored in liquid nitrogen for extended period will not change with time although laboratory trials with another strain have shown that its immunogenicity and infectivity were not affected by 15 years of continuous storage in liquid nitrogen (Bock et al. 1992).

2. Immunologic changes in field strains which will allow these strains to circumvent the protective immunity provided by the vaccine. This phenomenon has not been described in strains of *Babesia* but Walliker (1989) has described how genetic exchange in malaria parasites can result in selection of parasites expressing alternative antigen forms if these parasites are put under selection pressure by vaccines. As all strains tested to date have few alleles in common and appear to be capable of rapid genomic change (Dalrymple and Wright 1991, pers. comm.), selection pressure is considered to be the likely reason for the failures of the *B. bovis* vaccine from 1989–1990. Work is currently underway to test the hypothesis that field strains of *B. bovis* will show genomic and antigenic changes if passaged in previously vaccinated cattle. Work is also underway to catalogue breakthrough and other field strains of *B. bovis* with the use of DNA probes (Dalrymple and Wright 1991, pers. comm.). If the hypothesis proves to be correct, it implies that antigenically different strains must be incorporated in the vaccine at regular intervals, e.g. every 3–5 years, to ensure its continual efficacy (Bock et al. 1992).

References

Anonymous (1984) Tick and tick-borne disease control – a practical field manual. II. Tick-borne disease control. Food and Agricultural Organisation of the United Nations, Rome, pp 300–621

Bock RE, de Vos AJ, Kingston TG, Shiels IA, Dalgliesh RJ (1992) Investigations of break downs against babesiosis by living *Babesia bovis* vaccine. Vet Parasitol (in press)

Callow LL (1977) Vaccination against bovine babesiosis. In: Miller LH, Pino JA, McKelvey JJ (eds) Immunity to blood parasites of animals and man. Plenum, New York, pp 121–149

Callow LL (1978) Ticks and tick-borne diseases as a barrier to the introduction of exotic cattle to the tropics. World Anim Rev 28:20–25

Callow LL, Dalgliesh RJ (1980) The development of effective, safe vaccination against babesiosis and anaplasmosis in Australia. In: Johnston LAY, Cooper MG (eds) Ticks and tick-borne diseases. Australian Veterinary Association, Sydney, pp 4–8

Callow LL, Quiroga QC, McCosker PJ (1976) Serological comparison of Australian and South American strains of *Babesia argentina* and *Anaplasma marginale*. Int J Parasitol 6:307–310

Carson CA, Timms P, Cowman AF, Stewart NP (1990) *Babesia bovis*: evidence for selection of subpopulations during attenuation. Exp Parasitol 70:404–410

Dalgliesh RJ (1972) Theoretical and practical aspects of freezing parasitic protozoa. Aust Vet J 48:233–239

Dalgliesh RJ (1985) Practical considerations in the production and distribution of *Babesia bovis* vaccine. In: Irvin AD (ed) Immunization against theileriosis in Africa. ILRAD, Nairobi, Kenya, pp 125–128

Dalgliesh RJ, Callow LL, Mellors LT, McGregor W (1981) Development of a highly infective *Babesia bigemina* vaccine of reduced virulence. Aust Vet J 57:8–11

Dalgliesh RJ, Jorgensen WK, de Vos AJ (1990) Australian frozen vaccines for the control of babesiosis and anaplasmosis in cattle – a review. Trop Anim Health Prod 22:44–52

de Vos AJ (1979) Epidemiology and control of bovine babesiosis in South Africa. J S Afr Vet Assoc 50:357–362

de Vos AJ, Potgieter FT (1983) The effect of tick control on the epidemiology of bovine babesiosis. Onderstepoort J Vet Res 50:3–5

de Vos AJ, Potgieter FT (1991) Bovine babesiosis. In: Coetzer JAW, Kriek NPJ, Thomson GR, Tustin RC (eds) Infectious diseases of livestock in southern Africa. Oxford University Press, London (in press)

de Vos AJ, Bessenger R, Fourie CG (1982a) Virulence and heterologous strain immunity of South African and Australian *Babesia bovis* strains with reduced pathogenicity. Onderstepoort J Vet Res 49:133–136
de Vos AJ, Combrink MP, Bessenger R (1982b) *Babesia bigemina* vaccine: comparison of the efficacy and safety of Australian and South African strains under experimental conditions in South Africa. Onderstepoort J Vet Res 49:155–158
de Vos AJ, Dalgliesh RJ, Callow LL (1987) Babesia. In: Soulsby EJL (ed) Immune responses in parasitic infections: immunology, immunopathology and immunoprophylaxis. 3. Protozoa. CRC, Boca Raton, Florida, pp 183–222
Friedhoff KT (1987) Transmission of *Babesia* In: Ristic M (ed) Babesiosis of domestic animals and man. CRC, Boca Raton, Florida, pp 23–52
Gray JS, de Vos AJ (1981) Studies on a bovine *Babesia* transmitted by *Hyalomma marginatum rufipes* Koch, 1844. Onderstepoort J Vet Res 48:215–223
Jorgensen WK, de Vos AJ, Dalgliesh RJ (1989a) Infectivity of cryopreserved *Babesia bovis, Babesia bigemina* and *Anaplasma centrale* for cattle after thawing, dilution and incubation at 30 °C. Vet Parasitol 31:243–251
Jorgensen WK, de Vos AJ, Dalgliesh RJ (1989b) Comparison of immunogenicity and virulence between *Babesia bigemina* parasites from continuous culture and from a splenectomised calf. Aust Vet J 61:371–372
Lawrence JA, Norval RAI (1979) A history of ticks and tick-borne diseases of cattle in Rhodesia. Rhod Vet J 10:28–40
Lawrence JA, de Vos AJ, Irvin AD (1991) Theileriosis. In: Coetzer JAW, Kriek NPJ, Thomson GR, Tustin RC (eds) Infectious diseases of livestock in southern Africa. Oxford University Press, London (in press)
Levy MG, Ristic M (1980) *Babesia bovis*: continuous cultivation in a microaerophilus stationary phase culture. Science 207:1218–1220
Mahoney DF, Ross DR (1972) Epizootiological factors in the control of bovine babesiosis. Aust Vet J 48:292–298
Mahoney DF, Wright IG, Goodger BV (1979) Immunity in cattle to *Babesia bovis* after single infections with parasites of various origin. Aust Vet J 55:10–12
Mahoney DF, Wright IG, Goodger BV, Mirre GB, Sutherst RW, Utech KBW (1981) The transmission of *Babesia bovis* in herds of European and Zebu × European cattle infested with the tick, *Boophilus microplus*. Aust Vet J 57: 461–469
Mahoney DF, Wright IG, Goodger BV (1984) Immunisation against babesiosis: current studies and future outlook. Prev Vet Med 2:401–408
McCosker PJ (1981) The global importance of babesiosis. In: Ristic M, Kreier JP (eds) Babesiosis. Academic Press, New York, pp 1–24
Mellors LT, Dalgliesh RJ, Timms P, Rodwell BJ, Callow LL (1982) Preparation and laboratory testing of frozen vaccine containing *Babesia bovis, Babesia bigemina* and *Anaplasma centrale*. Res Vet Sci 32:194–197
Montenegro-James S, Kakoma I, Ristic M (1989) Culture-derived *Babesia* exoantigens as immunogens. In: Wright IG (ed) Veterinary protozoan and hemoparasite vaccines. CRC Press, Boca Raton, Florida, pp 61–97
Norval RAI (1983) Arguments against intensive dipping. Zimbabwe Vet J 14:19–25
Norval RAI, Short NJ (1984) Interspecific competition between *Boophilus decoloratus* and *Boophilus microplus* in southern Africa. Acarology 6, 2:1242–1246
Norval RAI, Fivaz BH, Lawrence JA, Daillecourt T (1983) Epidemiology of tick-borne diseases of cattle in Zimbabwe. I. Babesiosis. Trop Anim Health Prod 15:87–94
Pipano E, Hadani A (1984) Control of bovine babesiosis. In: Ristic M, Ambriose-Thomas P, Kreier JP (eds) Malaria and babesiosis. Martinus Nijhoff, Dordrecht, Netherlands, pp 263–303
Potgieter FT (1977) The life-cycle of *Babesia bovis* and *Babesia Bigemina* in ticks and in cattle in South Africa. PhD Thesis, Rand Afrikaans University
Rogers RJ, Dimmock CK, de Vos AJ, Rodwell BJ (1988) Bovine leucosis virus contamination of a vaccine produced in vivo against bovine babesiosis and anaplasmosis. Aust Vet J 65:285–287
Stewart NP, de Vos AJ, McHardy N, Standfast NF (1990) Elimination of *Theileria buffeli* infections from cattle by concurrent treatment with buparvaquone and primaquine phosphate. Trop Anim Health Prod 22:116–122

Sutherst RW, Maywald GF (1987) Modelling tick management. 2. The role of CLIMEX. In: Sutherst RW (ed) Ticks and tick-borne diseases. ACIAR Proc No 17, Australian Centre for International Agricultural Research, Canberra, Australia, pp 137–140

Timms P, Stewart NP (1989) Growth of *Babesia bovis* parasites in stationary and suspension cultures and their use in experimental vaccination of cattle. Res Vet Sci 47:309–314

Timms P, Dalgliesh RJ, Barry DN, Dimmock CK, Rodwell BJ (1983) *Babesia bovis*: comparison of culture-derived parasites, non-living antigen and conventional vaccine in the protection of cattle against heterologous challenge. Aust Vet J 60:75–77

Timms P, Stewart NP, de Vos AJ (1990) Study of virulence and vector transmission of *Babesia bovis* by use of cloned parasite lines. Infect Immun 58:2171–2176

Vega CA, Buening GM, Green TJ, Carson CA (1985) In vitro cultivation of *Babesia bigemina*. Am J Vet Res 46:416

Walliker D (1989) Implications of genetic exchange in the study of protozoan infections. Parasitology 99:S49–S58

Weilgama DJ, Jorgensen WK, Dalgliesh RJ, Navaratne M, Weerasinghe C (1989) Comparison between Sri Lankan and Australian strains of *Babesia bovis* in the vaccination of imported cattle in Sri Lanka. Trop Anim Health Prod 21:141–145

Wright IG, Goodger BV, Mahoney DF (1980) The irradiation of *Babesia bovis*. I. The difference in pathogenicity between irradiated and non-irradiated populations. Z Parasitol 63:47–57

Yunker CE, Kuttler KL, Johnson LW (1987) Attenuation of *Babesia bovis* by in vitro cultivation. Vet Parasitol 24:7–13

Chemical Control of Ticks on Cattle

J. Schröder[1]

Summary

Chemicals are an important part of efforts to control ticks on cattle. They can be grouped by their origin or chemical structure as plant extracts, organochlorines, organophosphates, carbamates, formamidines, synthetic pyrethroids, and macrocyclic lactones. Their efficacy has been proven by means of a variety of in vitro and in vivo test methods. Topical application of aqueous formulations by means of plunging or spraying has been the principal means of treatment for many years. Innovative, labour-saving formulations, such as pour-ons, injections and impregnated devices, have been developed over the last few years. As is the case with most chemicals used for pest control, ticks that are resistant to the majority of the different classes of acaricides, have been described from various parts of the world. Although *Boophilus* spp. are the prime culprits, multi-host ticks also display this ability. The benefits that are derived from chemical control of the harmful effects of ticks, can be offset by other factors, such as toxicity of the chemical, production losses from handling of the animals, and the direct cost of the chemical.

1 Introduction

Chemicals must be seen as an essential part of an integrated approach to tick control, which also includes host resistance, cattle movement, habitat manipulation, and vaccination. They should be used prudently, to optimise the cost-benefit ratio for tick control in the short term, whilst the long-term consideration should be to maximize the life of the acaricides, by delaying the development of resistance and preserving new chemicals for future use (Sutherst 1987). This is often easier said than done, because integrated control methods that are successful against single-host ticks in countries like Australia, are confounded by multi-host ticks in Africa (Schröder and Van Schalkwyk 1989).

This necessitates the use of chemicals for the control of ticks (Dorn et al. 1982; Rechav 1987), especially where European breeds rather than zebu or crossbred cattle are reared because of retailer preference (Norton 1987). But a farmer can also become "locked into" the use of acaricides after incurring the cost of constructing a dip tank, for instance, or eradicating locally occurring tick-borne diseases, with the concurrent risk of an epidemic if tick control is relaxed (Norton 1987; Schröder 1989, unpubl. observ.).

---

[1]Bayer (SA) Animal Health, PO Box 143, 1600 Isando, Republic of South Africa

2 Chemical Classes of Acaricides

Prior to 1940, various agents, such as botanical pesticides (e.g. pyrethrum and rotenone), and arsenical compounds were used for tick control (Ivie and Rowe 1986). Some, such as Cape aloes (Bedford and Wilken-Jorden 1934), were tested but found to be ineffective (recent preliminary findings suggest that there may yet be some substance in persistent anecdotal reports of aloes' efficacy against ticks (Schröder 1988, unpubl. observ.).

Veterinary ectoparasiticides are usually offshoots of synthesis and screening efforts aimed primarily at crop protection. Chlorinated hydrocarbons have fallen into disfavour because of their biological and environmental stability. By contrast, the organophosphates and carbamates, which affect acetylcholinesterase and were developed during the 1950s, are much less persistent in the environment, are highly biodegradable, and have more desirable tissue residue characteristics (Ivie and Rowe 1986).

Synthetic pyrethroids (which cause hyperexcitation, convulsions and death in arthropods), formamidines (whose mode of action has not been clearly defined), and avermectins (which act as GABA agonists in arthropod neurotransmission) have reached the market during the past decade or so (Hart 1986; Ivie and Rowe 1986). Chitin inhibitors and juvenile hormone analogues have not yet been used for tick control (Ivie and Rowe 1986).

2.1 Test Methods for Acaricidal Chemicals

The efficacy of chemicals against ticks can be determined by various means. The choice of a test method depends, apart from the preference of the investigator and the tradition of the laboratory doing the testing, on a variety of factors. The purpose of the test, the type of formulation and the mode of action are but a few considerations when a test method is selected.

A variety of in vitro (e.g. packet and sandwich larvae tests, engorged female immersion test, disposable pipette test, teabag test, and filter paper residue test) and in vivo methods described by various authors is listed by Drummond (1986). Larval tests (Shaw 1966; Luguru et al. 1984) may be more appropriate for resistance surveys, while the engorged female immersion test (Drummond et al. 1973) is better suited for efficacy screening (Stendel 1980). Barnard et al. (1981) used 3-month-old nymphs of *Amblyomma americanum* in a pipette method to establish baseline susceptibility data for a range of chemicals.

Larval bioassay tests as indicators of adult tick resistance have limited validity, because adults are less susceptible than larvae of the same strain (Solomon et al. 1979; Schröder and Van Schalkwyk 1989). On the other hand, the engorged female immersion test uses a tick that is no longer parasitic, and cannot determine residual activity. Few of the in vitro tests can evaluate non-topical or systemic formulations (Schröder 1987).

The most accurate indicator of the effective concentration of an acaricide is the in vivo test (Stendel 1980), but it is also the most expensive and time consuming (Drummond 1986). In vivo testing of the intended final formulation has been the requirement for proving the efficacy of a chemical acaricidel for registration in South Africa for a number of years (Anonymous 1977) and has been published as a Standard Method of the South African Bureau of Standards (Anonymous 1987).

Innovative modes of application necessitate new test methods, such as fluorescent and other techniques to demonstrate the dermal distribution of the active ingredient from a pour-on formulation or impregnated devices (Hamel and Van Amelsfoort 1986; Stendel 1986; Taylor et al. 1987).

2.2 Reports on the Efficacy of Acaricides

There are many publications describing the acaricidal efficacy of various chemicals. The list cited here is quite long, by no means complete, and starts with arsenic (Bedford 1929). Other chemicals mentioned represent plant extracts (Bedford and Wilken-Jorden 1934), amidines (Haigh and Gichang 1980), macrocyclic lactones (in the form of ivermectin) (Nolan et al. 1981; Pegram and Lemche 1985; Schröder et al. 1985; Cramer et al. 1988b), the organophosphates, organochlorines, and synthetic pyrethroids.

Among the organophosphates are propetamphos (Anonymous 1981), chlorfenvinphos and dioxathion (Baker and Thompson 1966; Hammant 1977), coumaphos, malathion, ronnel, and stirofos (Barnard and Jones 1981), chlorpyrifos, crotoxyphos, phosmet, carbophenothion, ethion, phosphamidon, crufomate, bromophos-ethyl, trichlorfon, diazinon, fenthion, and famphur (Drummond 1981). Examples of organochlorines are toxaphene (Baker and Thompson 1966), lindane and DDT (Drummond 1981). Synthetic pyrethroids are represented by flumethrin (Dorn et al. 1982; Hamel et al. 1982; Hopkins and Woodley 1982; Stendel and Fuchs 1982; Lemche and Pegram 1987), cyhalothrin (Stubs et al. 1982), cypermethrin (Taylor and Elliott 1987) and alphamethrin (Schröder and Van Schalkwyk 1989).

The ticks against which the efficacy of these chemicals has been reported include *Amblyomma americanum* (Barnard and Jones 1981), *Amblyomma hebraeum, Rhipicephalus appendiculatus, Rhipicephalus evertsi evertsi* (Baker and Thompson 1966; Dorn et al. 1982), *Boophilus decoloratus* (Baker and Thompson 1966), *Amblyomma cajennense* (Drummond 1981), *Boophilus microplus* (Hamel et al. 1982; Hopkins and Woodley 1982), and *Ixodes ricinus* (Taylor and Elliott 1987).

The dosage of active ingredient per animal (either given systemically, or extrapolated from the recommended concentration for topical formulations) varies according to the chemical and the formulation: 25 mg/kg for camphechlor, 2.5 mg/kg for amitraz, 1.0–5.0 mg/kg for organophosphates, 0.3–2.0 mg/kg for synthetic pyrethroids, and 0.02–1.0 mg/kg for ivermectin (see Table 1).

There is a great deal of support for establishing dosage-mortality data for a new chemical, before resistance develops (Barnard et al. 1981). In southern

Table 1. Summary of the dosage levels of some acaricides described in the literature

Active ingredient	Recommended concentration (mg/l, or ppm)	Dosage (mg/kg live mass)[a]	Author
Inorganics, plant extracts			
Arsenic			Bedford (1929)
	1600–3200	16–32	Matthewson and Baker (1975)
Aloes	–	65–133 (once) 63–128/d (5d)	Bedford and Wilken-Jorden (1934)
Organophosphates			
Chlorfenvinphos	100, 200 and 500	1–5	Baker and Thompson (1966)
Propetamphos			Anonymous (1981)
Dioxathion	1500	15	Barnard and Jones (1981)
	500	5	Baker and Thompson (1966)
	500	5	Tatchell et al. (1986)
Coumaphos			Roulston et al. (1968)
	1250	12.5	Barnard and Jones (1981)
Stirofos	3500–5000	35–50	Barnard and Jones (1981)
Malathion	5000–10000	50–100	Barnard and Jones (1981)
Ronnel	7500	75	Barnard and Jones (1981)
Organochlorines			
Camphechlor	2500	25	Baker et al. (1966)
Toxaphene	3000	30	Barnard and Jones (1981)
Lindane	300–600	3–6	Barnard and Jones (1981)
Formamidines			
Amitraz	250	2.5	Coetzee et al. (1987a)
Amitraz			Haigh and Gichang (1980)
Synthetic pyrethroids			
Flumethrin	30, 50, 75	0.3–0.75	Dorn et al. (1982)
	–	0.5–1.5	Dorn et al. (1986); Dorn and Pulga (1985)
	15, 30, 60	0.15–0.6	Hamel et al. (1982)
	–	1	Hamel and Duncan (1986)
	–	1	Hamel and Van Amelsfoort (1985)
	30–75	0.3–0.75	Hopkins and Woodley (1982)
	–	1	Hopkins et al. (1985)
			Lemche and Pegram (1987)
	–	0.5–1.5	Sosa (1985)
	30, 40	0.3–0.4	Stendel and Fuchs (1982)
	50	0.5	Coetzee et al. (1987a)
Cyhalothrin			Stubs et al. (1982)
Deltamethrin	50	0.5	Coetzee et al. (1987a)
Fenvalerate	200	2	Coetzee et al. (1987b)
Macrocyclic lactones			
Ivermectin	–	0.2–1.0	Cramer et al. (1988a)
	–	0.2	Cramer et al. (1988b)
	–	0.2	Schröder et al. (1985)
	–	0.02–0.06	Soll et al. (1987)

[a] Where a topical (dip or spray) formulation is described, the dosage per kg live mass was calculated on the following assumptions: animal weighing 350 kg, carrying out 3.5 l dip wash.

Africa, such data will have to be generated for all the major tick species, preferably against standardized isolates, but definitely utilizing the same test or battery of tests.

3 Innovative Formulations of Acaricidal Chemicals

While the discovery and development of new molecules remains a prime objective, a further challenge facing the chemical/pharmaceutical industry in the field of chemical tick control is to develop new formulations. We have, alas, not progressed very far beyond the days at the turn of the century when desperate farmers plunged their tick-infested cattle bodily into vats filled with toxicant. Spraying dip wash onto the animals, either mechanically in a spray race, or manually by means of a knapsack sprayer for example, represented an innovation in application method, but the formulations essentially remained unchanged.

It is during the last decade and a half that the greatest innovations in formulations have been made. Many impregnated slow-release devices for topical application have been described. These include ear tags containing stirifos, chlorpyrifos, ronnel, fenvalerate, alphamethrin, dichlorvos, cypermethrin, flucythrinate, propetamphos, and fluvalinate (Gladney 1976; Ahrens et al. 1977; Ahrens and Cocke 1978; Taylor et al. 1984; Rechav 1987; Schröder and Van Schalkwyk 1989), neck bands and horn bands with propoxur and stirifos (Gladney 1976; Ahrens et al. 1977), and tail bands (Taylor et al. 1987; Schröder and Van Schalkwyk 1989).

Ear tags have been tested against *Amblyomma maculatum* (Gladney 1976; Ahrens et al. 1977; Ahrens and Cocke 1978), *Amblyomma hebraeum, Boophilus decoloratus, Rhipicephalus evertsi evertsi* (Schröder and Van Schalkwyk 1989), and *Rhipicephalus appendiculatus* (Rechav 1987; Schröder and Van Schalkwyk 1989).

Pour-ons, which act topically (and which contain mainly synthetic pyrethroids) (Hamel 1984; Dorn and Pulga 1985; Hamel and Van Amelsfoort 1985; Hopkins et al. 1985; Sosa 1985; Stendel 1985; Dorn et al. 1986; Hamel and Duncan 1986; Taylor and Elliott 1987; Ahrens et al. 1988) or systemically (with ivermectin, or an organophosphate like phosmet) (Pitman and Rostas 1981; Cramer et al. 1988b) and other systemic (some of them slow-release) formulations offer additional means of applying chemicals for tick control. These include an ivermectin-containing subcutaneous injection (Cramer et al. 1988a), oral capsule (Drummond et al. 1981) and osmotic pump (Soll et al. 1987), and a sustained-release bolus containing famphur (Hair et al. 1979).

Much of the cost of acaricidal treatment of cattle is made up of the difficult to measure, but instinctively and universally, accepted, production losses caused by the handling of the animals during mustering, treatment and subsequent dispersal. This has been the driving force behind the quest for easy-to-apply, preferably long-acting (and therefore labour-saving) formulations.

New formulations (e.g. pyrethroid impregnated ear tags and pour-ons) are more expensive than conventional dips and sprays, but have less expenditure on hidden costs such as capital, facilities, maintenance, and staff (Pegram and Chizyuka 1987).

4 Resistance to Acaricides

According to our current understanding of the development of resistance, it is only a matter of time before resistance to a new chemical is established because somewhere there is at least one individual tick that is genetically tolerant to the new toxicant. Exposure to the chemical will be the selection pressure required to ensure that the individual's offspring constitute an ever-increasing percentage of the tick population, leading eventually to the establishment of 'resistance' to that chemical.

This process can evidently be quite rapid, although the 18 months reported by Coetzee et al. (1987b) is probably open to question. They mention that circumstantial evidence of DDT-cross-resistance exists on the property in question. Their doubt about the longevity of pyrethroids in the market seems ill-founded when one considers that this group of chemicals has been used in South Africa for the control of ticks since 1979.

Few of the chemical acaricides known today have escaped the phenomenon of resistance. In Africa, ticks that are resistant to arsenic (Du Toit et al. 1941; Matthewson and Baker 1975), organochlorines (Baker et al. 1977; Solomon et al 1979; Lourens 1980; Coetzee et al. 1987a), organophosphates (Solomon et al. 1979), carbamates (Roulston et al. 1968), and synthetic pyrethroids (Coetzee et al. 1987b) have been described.

Since the first description of resistance of *Boophilus decoloratus* to acaricides in southern Africa (Du Toit et al. 1941), five types of resistance have been defined, viz. arsenic, DDT-pyrethrum, organochlorine, organophosphorus-carbamate, and amidine (Nolan and Schnitzerling 1986). Fortunately, the complexity and spectrum of resistance shown by *Boophilus* (Roulston et al. 1968; Baker et al. 1979; Hopkins and Woodley 1982; Coetzee et al. 1987a; Bull and Ahrens 1988; Harris et al. 1988) has not been encountered in other ectoparasites (Nolan and Schnitzerling 1986), but several species of multi-host ticks have yielded isolates that are resistant to chemicals (Matthewson and Baker 1975; Baker et al. 1977; Lourens 1979; Solomon et al. 1979; Lourens 1980)

4.1 Chemical Acaricides as Pollutants

Few people have problems taking proper care when handling and disposing of obviously toxic substances like arsenic or the organophosphates. It was the discovery that evidently safe compounds, like ivermectin, can have a profound impact on the environment (Wall and Strong 1987), that made people once again re-assess the implications of the use of chemicals in the control of ticks.

5 Benefits of Chemical Tick Control

Tick control through the use of ivermectin resulted in increased live mass gain in cattle in Zambia (Pegram and Lemche 1985). Several other papers mention the benefits to be derived from chemical control of ticks (Drummond et al. 1988; Morzaria et al. 1988), but it appears that such reports have to be accepted with reservation.

The need for intensive chemical tick control in semi-arid areas of Africa is questioned by Tatchell et al. (1986), who failed to demonstrate an improvement in live mass gain or decrease in mortality from tick-borne diseases in cattle that were treated intensively. They point out, however, that the tick numbers were low, they dealt with indigenous (relatively tick-resistant) cattle, and the greatest causes of production losses during their trial were drought and malnutrition.

The low production response to frequent dipping of cattle that is seen in Africa could possibly be attributed to chronic toxicity. The growth of Zebu cattle has been shown to be affected by toxicity of organophosphates in Australia (Sutherst 1987). Zebu cattle are perhaps also susceptible to amidine chemicals. Ideally, the effects of acaricides on animal growth in the absence of ticks need to be known, so that the benefit to be derived from their use under different conditions and tick challenges can be calculated (Sutherst and Kerr 1987). Some acaricides are also toxic to other parasites (e.g. flies, mosquitoes, and, in the case of systemic chemicals, also helminths). This complicates the estimation of the benefits of tick control (Sutherst and Kerr 1987).

The total cost of chemical tick control is made up of the cost of the chemical, the handling of the livestock, the establishment of resistance (which will increase future costs), and the cost of labour and capital allocated to tick control (Sutherst 1987).

References

Ahrens EH, Cocke J (1978) Comparative test with insecticide-impregnated ear tags against the Gulf Coast tick. J Econ Entomol 71:764–765

Ahrens EH, Gladney WJ, McWhorter GM, Deer JA (1977) Prevention of screwworm infestation in cattle by controlling Gulf Coast ticks with slow release insecticide devices. J Econ Entomol 70:581–585

Ahrens EH, Davey RB, George JE, Cooksey LM (1988) Flumethrin applied as a pour-on and whole-body spray for controlling cattle tick (Acari: Ixodidae) on cattle. J Econ Entomol 81:1133–1136

Anonymous (1977) A guide to the requirements for the registration of pesticides for the control of ectoparasites. Draft 2. Veterinary Research Institute, Onderstepoort

Anonymous (1981) Propetamphos for the control of ticks. Livest Int 9:8

Anonymous (1987) Standard method 1096–1987. Stock remedies: efficacy confirmation of tickicides on cattle. Gov Gaz 10881, 4 September:13

Baker JAF, Thompson GE (1966) Supona for cattle tick control 1. Handspraying trials. J S Afr Vet Med Assoc 37:367–372

Baker JAF, Thompson GE, Miles JO (1977) Resistance to toxaphene by the bont tick, *Amblyomma hebraeum*. J S Afr Vet Assoc 48:59–65

Baker JAF, Jordaan JO, Robertson W (1979) Ixodicidal resistance in *Boophilus decoloratus* (Canestrini) in the Republic of South Africa and Transkei. J S Afr Vet Assoc 50:296–301

Barnard DR, Jones BG (1981) Field efficacy of acaricides for control of the Lone Star tick on cattle in southeastern Oklahoma. J Econ Entomol 74:558–560
Barnard DR, Jones BG, Rogers GD, Mount GA (1981) Acaricide susceptibility in the Lone Star tick: assay techniques and baseline data. J Econ Entomol 74:466–469
Bedford GAH (1929) The effect upon ticks of dipping cattle regularly at short intervals in arsenical baths. 15th Rep Dir Vet Serv Anim Ind:551–573
Bedford GAH, Wilken-Jorden TJ (1934) Researches into dips and dipping. C. Miscellaneous. The effect of dosing aloes to tick-infected cattle. Onderstepoort J Vet Sci Anim Ind 11:267–269
Bull DL, Ahrens EH (1988) Metabolism of coumaphos in susceptible and resistant strains of *Boophilus microplus*. J Med Entomol 25:94–98
Coetzee BB, Stanford GD, Davis DAT (1987a) The resistance spectrum shown by a fenvalerate-resistant strain of blue tick (*Boophilus decoloratus*) to a range of ixodicides. Onderstepoort J Vet Res 54:79–82
Coetzee BB, Stanford GD, Davis DAT (1987b) Resistance by the blue tick (*Boophilus decoloratus*) to the synthetic pyrethroid, fenvalerate. Onderstepoort J Vet Res 54:83–86
Cramer LG, Bridi AA, Amaral NK, Gross SJ (1988a) Persistent activity of injectable ivermectin in the control of the cattle tick *Boophilus microplus*. Vet Rec 122:611–612
Cramer LG, Carvalho LAF, Bridi AA, Amaral NK, Barrick RA (1988b) Efficacy of topically applied ivermectin against *Boophilus microplus* (Canestrini 1887) in cattle. Vet Parasitol 29:341–349
Dorn H, Pulga M (1985) Field trials with flumethrin pour-on against *Boophilus microplus* in Brazil. Vet Med Rev 2:146–151
Dorn H, Hamel HD, Stendel W (1982) The efficacy of flumethrin (Bayticol) against multihost cattle ticks in South Africa under field conditions. Vet Med Rev 2:147–157
Dorn H, Romano A, Pulga M, Rodriguez O (1986) Field trials with flumethrin pour-on in Argentina, Brazil and Colombia against *Boophilus microplus* – a new method of tick control in cattle. Vet Med Rev 1:40–45
Drummond RO (1981) Susceptibility of the Cayenne tick to acaricides. J Econ Entomol 74:470–472
Drummond RO (1986) Acarine infestation of domestic animals. In: Campbell WC, Rew RS (eds) Chemotherapy of parasitic diseases. Plenum, New York, pp 567–583
Drummond RO, Ernst SE, Trevino JL, Gladney WJ, Graham OH (1973) *Boophilus annulatus* and *Boophilus microplus*: laboratory tests of insecticides. J Econ Entomol 66:130–133
Drummond RO, Whetstone TM, Miller JA (1981) Control of ticks systemically with Merck MK-933, an avermectin. J Econ Entomol 74:432–436
Drummond RO, George JE, Kunz SE (1988) Control of arthropod pests of livestock: a review of technology. CRC, Boca Raton, Florida, pp 12–14
Du Toit R, Graf H, Bekker PM (1941) Resistance to arsenic as displayed by the single-host blue tick *Boophilus decoloratus* (Koch) in a localised area of the Union of South Africa: Preliminary report. J S Afr Vet Med Assoc 12:50–58
Gladney WJ (1976) Field trials of insecticides in controlled-release devices for control of the Gulf Coast Tick and prevention of screwworm in cattle. J Econ Entomol 69:757–760
Haigh AJB, Gichang MM (1980) The activity of amitraz against infestation of *Rhipicephalus appendiculatus*. Pest Sci 11:674–678
Hair JA, Gladney WJ, Davey RB, Drummond RO, Teel PD (1979) Sustained-release famphur bolus for control of *Boophilus* ticks. J Econ Entomol 72:135–138
Hamel HD (1984) A new method of tick control in cattle. Proc XIIIth Wld Congr Dis Cattle, Durban, S A, Sept 17–21, 1:448–451
Hamel HD, Duncan IM (1986) Cattle tick control in Zimbabwe with flumethrin 1% pour-on. Vet Med Rev 2:115–122
Hamel HD, Van Amelsfoort AF (1985) Tick control with flumethrin 1% m/n pour-on under South African field conditions. Vet Med Rev 2:132–145
Hamel HD, Van Amelsfoort AF (1986) Demonstration of dermal distribution of flumethrin 1% by means of a fluorescence technique. Vet Med Rev 1:34–39
Hamel HD, Esteves W, Hees B, Pulga M, Roessger W (1982) Field trials with Bayticol against *Boophilus microplus* in Brazil. Vet Med Rev 2:140–146
Hammant CA (1977) The introduction of dioxathion for cattle tick control in the tribal trust lands of Rhodesia. Rhod Vet J 8:67–70

Harris RL, George JE, Ahrens EH, Davey RB, Bazan HO (1988) Selection for resistance to coumaphos in a strain of southern cattle tick. J Econ Entomol 81:545–548
Hart RJ (1986) Mode of action of agents used against arthropod parasites. In: Campbell WC, Rew RS (eds) Chemotherapy of parasitic diseases. Plenum, New York, pp 585–601
Hopkins TJ, Woodley IR (1982) The efficacy of flumethrin against susceptible and organophosphate resistant strains of the cattle tick *Boophilus microplus* in Australia. Vet Med Rev 2:130–139
Hopkins TJ, Woodley IR, Blackwell R (1985) The safety and efficacy of flumethrin pour-on used to control *Boophilus microplus* on cattle in Australia. Vet Med Rev 2:112–125
Ivie GW, Rowe LD (1986) Chemistry of drugs used against arthropod parasites. In: Campbell WC, Rew RS (eds) Chemotherapy of parasitic diseases. Plenum, New York, pp 507–529
Lemche J, Pegram RG (1987) Control of cattle ticks using flumethrin in central Zambia. Vet Rec 121:110–111
Lourens JHM (1979) Susceptibility of a dieldrin-resistant and susceptible strain of the tick *Rhipicephalus evertsi evertsi* from East Africa to carbamate and OP acaricides. Pest Sci 10:495–501
Lourens JHM (1980) Inheritance of organochlorine resistance in the cattle tick *Rhipicephalus appendiculatus* in East Africa. Bull Entomol Res 70:1–10
Luguru SM, Banda DS, Pegram RG (1984) Susceptibility of ticks to acaricides in Zambia. Trop Anim Health Prod 16:21–26
Matthewson MD, Baker JAF (1975) Arsenic resistance in species of multi-host ticks in the RSA and Swaziland. J S Afr Vet Assoc 46:341–344
Morzaria SP, Irvin AD, Wathanga J, D'Souza D, Katende J, Young AS, Scott J, Gattinby G (1988) The effect of East Coast fever immunisation and different acaricidal treatments on the productivity of beef cattle. Vet Rec 123:313–320
Nolan J, Schnitzerling HJ (1986) Drug resistance in arthropod parasites. In: Campbell WC, Rew RS (eds) Chemotherapy of parasitic diseases. Plenum, New York, pp 603–620
Nolan J, Schnitzerling HJ, Bird P (1981) Evaluation of the potential of systemic slow release chemical treatments for control of the cattle tick (*Boophilus microplus*) using ivermectin. Aust Vet J 57:493–497
Norton GA (1987) A Strategic approach for tick control programs. In: Sutherst RW (ed) Ticks and tick-borne diseases. Proceedings of an international workshop on the ecology of ticks and epidemiology of tick-borne diseases, held at Nyanga, Zimbabwe, 17–21 February 1986. ACIAR, Canberra, pp 126–132
Pegram RG, Chizyuka HGB (1987) Towards an assessment of the economic impact of ticks on rural development. In: Sutherst RW (ed) Ticks and tick-borne diseases. Proc Int Worksh Ecology of ticks and epidemiology of tick-borne diseases, held at Nyanga, Zimbabwe, 17–21 Feb 1986. ACIAR, Canberra, pp 104–107
Pegram RG, Lemche J (1985) Observations on the efficacy of ivermectin in the control of cattle ticks in Zambia. Vet Rec 117:551–554
Pitman IH, Rostas SJ (1981) Topical drug delivery to cattle and sheep. J Pharm Sci 70:1181–1194
Rechav Y (1987) Use of acaricide-impregnated ear tags for controlling the brown ear tick in South Africa. J Econ Entomol 80:822–825
Roulston WJ, Stone BF, Wilson JJ, White LI (1968) Chemical control of an organosphorus and carbamate resistant strain of *Boophilus microplus* from Queensland. Bull Entomol Res 58:379–392
Schnitzerling H, Nolan J, Hughes S (1989) Toxicology and metabolism of isomers of flumethrin in larvae of pyrethroid-susceptible and resistant strains of the cattle tick *Boophilus microplus*. Exp Appl Acarol 6:47–54
Schröder J (1987) Chemical control of the heartwater vectors. Onderstepoort J Vet Res 54:517–520
Schröder J, Van Schalkwyk PC (1989) Die doeltreffendheid van alfametriendeurweekte oorplaatjies teen beesbosluise. J S Afr Vet Assoc 60:79–82
Schröder J, Swan GE, Soll MD, Hotson IK (1985) Efficacy of ivermectin against ectoparasites of cattle in South Africa. J S Afr Vet Assoc 56:31–35
Shaw RD (1966) Culture of an organophosphorus-resistant strain of *Boophilus microplus* (Can.) and an assessment of its resistance spectrum. Bull Entomol Res 56:389–405
Soll MD, Carmichael IH, Gross SJ (1987) Control of induced infestations of adult *Amblyomma hebraeum* with sustained release ivermectin. Onderstepoort J Vet Res 54:17–20

Solomon KR, Baker MK, Heyne H, Van Kleef J (1979) The use of frequency diagrams in the survey of resistance to pesticides in ticks in southern Africa. Onderstepoort J Vet Res 46:171–177

Sosa E (1985) Evaluation of the efficacy and residual effect of flumethrin pour-on against *Boophilus microplus* in cattle in Uruguay. Vet Med Rev 2:126–131

Stendel W (1980) The relevance of different test methods for the evaluation of tick controlling substances. J S Afr Vet Assoc 51:147–152

Stendel W (1985) Experimental studies on the tickicidal effect of Bayticol pour-on. Vet Med Rev 2:99–111

Stendel W (1986) Studies on the distribution of flumethrin pour-on on the skin surface and the haircoat of cattle. Vet Med Rev 1:28–33

Stendel W, Fuchs R (1982) Laboratory evaluation of flumethrin a new synthetic pyrethroid for the control of one- and multi-host ticks. Vet Med Rev 2:115–129

Stubs VK, Wilshire C, Webbes LG (1982) Cyhalothrin – a novel acaricidal and insecticidal synthetic pyrethroid for the control of the cattle tick and the buffalo fly. Aust Vet J 59:152–155

Sutherst RW (1987) Modelling tick management. 1. Introduction. In: Sutherst RW (ed) Ticks and tick-borne diseases. Proc Int Worksh Ecology of ticks and epidemiology of tick-borne diseases, held at Nyanga, Zimbabwe, 17–21 Feb 1986. ACIAR, Canberra, pp 133–136

Sutherst RW, Kerr JD (1987) Losses in livestock productivity caused by ticks and tick-borne diseases. In: Sutherst RW (ed) Ticks and tick-borne diseases. Proc Int Worksh Ecology of ticks and epidemiology of tick-borne diseases, held at Nyanga, Zimbabwe, 17–21 Feb 1986. ACIAR, Canberra, pp 108–112

Tatchell RJ, Chimwani D, Chirchir SJ, Ong'are JO, Mwangi E, Rinkanya F, Whittington T (1986) A study of the justification for intensive tick control in Kenyan rangelands. Vet Rec 119:401–403

Taylor SM, Elliott CT (1987) Deltamethrin and flumethrin pour-on formulations for the control of *Ixodes ricinus* on cattle. Vet Rec 120:278

Taylor SM, Kenny J, Mallon TR, Elliott CT, McMurray C, Blanchflower J (1984) Efficacy of pyrethroid-impregnated ear tags for prophylaxis of tick borne diseases of cattle. Vet Rec 114:454–455

Taylor SM, Elliott CT, Blanchflower WJ (1987) A comparison of cypermethrin distribution on cattle hair after application of impregnated ear and tail tags. Pest Sci 21:39–43

Wall R, Strong L (1987) Environmental consequences of treating cattle with the antiparasitic drug ivermectin. Nature 327:418

Organism Index

Excepting Table 1 of Keirans

Subject Index

MIX
Papier aus verantwortungsvollen Quellen
Paper from responsible sources
FSC® C105338

If you have any concerns about our products,
you can contact us on
ProductSafety@springernature.com

In case Publisher is established outside the EU,
the EU authorized representative is:
Springer Nature Customer Service Center GmbH
Europaplatz 3, 69115 Heidelberg, Germany

Printed by Libri Plureos GmbH
in Hamburg, Germany